Transforming the

JAMES RIVER

IN RICHMOND

Transforming the

JAMES RIVER
IN RICHMOND

RALPH HAMBRICK

Foreword by Bill Street, Chief Executive Officer, James River Association

THE
History
PRESS

Published by The History Press
Charleston, SC
www.historypress.com

Front cover images: "Eagle on the James River" by Edward Episcopo, *courtesy of Scenic Virginia*; "Railroad Bridge at Sunset" by Bill Piper, *courtesy of Scenic Virginia;* rocks on the river, *courtesy of Scott Weaver;* rafting in downtown Richmond, *courtesy of Riverside Outfitters.*

Back cover image courtesy of Tricia Pearsall.

First published 2020

Manufactured in the United States

ISBN 9781467145350

Library of Congress Control Number: 2019956034

Dedicated to
Early James River advocates
and
All those who contribute to the health, beauty and enjoyment
of the James River in Richmond

CONTENTS

FOREWORD

The James River has many stories. Because many of them involve important historic events that shaped our country, Congress designated the James River as "America's Founding River." However, most of those stories are about what people did on or along the river. The story that has not been told as much is the story of the river itself.

Ralph Hambrick in *Transforming the James River in Richmond* tells the story of the river and what it means to Richmond, Virginia's capital city. He describes how the James River changed from one of the most polluted rivers in the nation to one of the most improved and loved rivers anywhere. That conversion and the effort propelling it is remarkable enough that it earned the James River the 2019 Thiess International Riverprize, the most prestigious and sought after award for management of a waterbody.

Because he was involved in many of them, Ralph is able to recount in great detail the many local events along with the personalities that contributed to that transformation in Richmond. His work is an appropriate celebration of those accomplishments and contributors, but it is perhaps even more important as a reminder of the level of commitment required to bring about change. As Ralph states many times throughout the book, there is still much work to be done. The story is not finished. So I hope that current and future river stewards will read this book to draw inspiration and to learn the lessons of the past to help in writing the next chapter of the story.

At a time when environmental issues have catapulted to the top on the list of public concerns, particularly for younger generations, it is important to have examples of success that provide hope and encouragement. The story of the James River is just that. We must use these examples of vision, courage and persistence to strengthen our resolve to do our part locally and globally. I encourage everyone who lives in our region to read this remarkable story of the James River and then jump in and help out!

—William H. Street
Chief Executive Officer
James River Association

PREFACE

A working title of this book in its early stages was "from sewer to park." Those words sound like hyperbole, but they are essentially accurate as a description of the seventy-year transformation of the James River in Richmond. For an extended period, the river was used as a sewer, and in a real sense, it is becoming a park. The formal James River Park System is not the total meaning of "park" in the former title, although it is a big part of it. The river as a whole in Richmond has moved toward a park-like quality, even though much of it is not contained within the dedicated James River Park System. At one level, the story is unique; at another, it is the story of urban rivers throughout the nation. It is unique in its particulars, but it is one of many rivers that has begun a journey toward restoration and a new pattern of use.

Traditionally, rivers were treated as utilitarian resources, or threats to be tamed, or both. They were used for transportation, drinking water, water power, irrigation and, yes, waste disposal. Although a means of transportation in one direction, they were a barrier in another. The floods they sometimes produced were a threat to be overcome. Rivers were considered imperfect tools, so canals, dams, bridges, floodwalls and pipes were installed. The James River in Richmond did not escape these efforts to make it more useful for human purposes, although it was not "engineered" as dramatically as some rivers around the country and the world. Multiple dams were built, canals dug, tributaries channeled, sewer and drainage pipes installed and factories and rail lines constructed on its banks.

By the mid-twentieth century, however, the industrial uses of the riverbanks had peaked and for the most part were abandoned. Richmond residents turned their backs to the river, treating it as the sewer it had become. Sometime around mid-century, the trends began to reverse, and the river began its journey from sewer to park, from a neglected and abused river to the centerpiece and pride of the city. That journey has not been uneventful; it has been a story of both conflict and cooperation, along with a large dose of passion.

Hopefully, the book offers a sense of the effort, struggle, successes and disappointments that have made the James River in Richmond the attraction that it is today. And hopefully the book captures a bit of the ambience of the river—a feel for the enjoyment the river provides.

ACKNOWLEDGEMENTS

I began dabbling with the idea for this book when Dr. R.B. Young, chair of the Falls of the James Scenic River Advisory Committee, asked me to write a brief history of the committee on the occasion of its thirtieth anniversary. That summary was made available to a nice crowd on Brown's Island in 2002. During presentations, a long train rumbled by on the trestle at the edge of the island, drowning out the speakers but not dampening the enjoyment of the event. Dr. Young, Louise Burke and John Pearsall retired from the Scenic River Board not long after the anniversary event, and each of them—as did Sue Cecil, who waited a while to retire—gave me access to several boxes of notes, correspondence, news clippings and meeting minutes. They also added to my education in a number of conversations about the river, its history and issues that needed to be addressed. Without their contributions, I would not have begun.

I was still on the faculty at Virginia Commonwealth University when I began, and I can say with confidence that I would have finished the book many years sooner had I not retired. There have been so many interesting things to do in retirement that I often put writing aside. I do wish to thank my dean at Virginia Commonwealth University, Stephen Gottfredson, who provided a partial reduction in teaching load one semester so that I could get started on the project.

Any honest author acknowledges that his or her work is not accomplished alone, and that certainly applies to this work. The list of those who contributed in one way or another is long. In addition to Burke, Cecil, Pearsall and Young, many others contributed time and knowledge to the project, including a new generation of members of the Scenic River Committee. Some read early drafts and offered feedback, some sat for interviews (or more accurately extended conversations), some contributed through interactions in meetings or events and some provided assistance in a variety of other ways (like chasing down photographs). My sincere thanks to them all: Janit Llewellyn Allen, Heather Barrar, Stuart Bateman, George Bruner, John Bryan, Nathan Burrell, Kimberly Conley, Lynn Crump, Alex Dahm, Will Daniel, Justin Doyle, Maureen Egan, Lorne Field, Greg Garman, Richard Gibbons, John Heerwald, Beverley Hundley, Patti Jackson, Mariane Jorgenson, Buzz Kraft, Lyn Lanier, James McCarthy, Katherine Mitchell, John Moeser, Mike Ostrander, Jack Pearsall, Tricia Pearsall, Charles Peters, Leighton Powell, Robert Steidel, Robert Stone, Bill Street, Bill Trout, Greg Velzy, Jennifer Wampler, Charles Ware, Alan Weaver, Ralph White, Bryce Wilk, Hanna Wolpert, Nancy Woodson, Anne Wright and Alicia Zatcoff.

A special thanks also goes to all those who contributed their photographs: Scott Adams, Edward Episcopo, Adam N. Goldsmith, Nancy Helms, Richard Kidd, Harold Lanna, Tricia Pearsall, Bill Piper, Scott Weaver, Don Wilson, Anne Wright, Rich Young and Leighton Powell, who arranged for use of photographs from Scenic Virginia's annual photo contests. Several organizations also contributed photographs that capture moments in the seventy-year transformation of the river: Cabell Library Archives of Virginia Commonwealth University, Historic Richmond Foundation, James River Association, Library of Congress, Richmond Department of Public Utilities, Riverside Outfitters, The Valentine and Venture Richmond. Thanks also are due to James River Park System for allowing adaptation of its map, Roger Sattler of Sattler Creative for preparing the map for this book, Richmond Camera for digitizing any number of prints and slides and Chuck Rudisill at Richmond Camera for patience with my many photography questions. And certainly I extend my appreciation to the professionals at The History Press for making the process as easy as it could be. My apologies to any whom I have inadvertently omitted.

Most assuredly, thanks to my wife, Linda, for putting up with multiple boxes of files scattered around the house, for reading a number of chapters, for enduring my absences while I was somewhere on or along the river, at a meeting or hidden away in my office. Her support was indispensable.

All of the above are absolved of any errors of commission or omission that remain.

1

TRANSFORMATION

INTRODUCING AN URBAN RIVER

For years, seemingly everyone has harped on the obvious: Richmond's riverfront is a jewel, and the city—region—ought to do more to capitalize on it. But transformation does not take place overnight.
—staff writer, Richmond Times-Dispatch, *February 7, 2014*

Many words might be used to characterize the change in this urban river over seventy years: *transformation, reclamation, modernization, restoration, renaissance.* These terms capture not simply the changes that have occurred in recent decades but also indicate the direction of that change. This is a paradigm shift from the utilitarian use of the river to the use of the river as an amenity—from water power for manufacturing and depository for waste to recreation and sightseeing.

Part of the good news is that the more environmentally healthy a river, the better it serves its new function as an amenity. In the old paradigm, use and environmental quality often were at odds; in the modern paradigm, use and environmental health are mutually reinforcing. Controlling the river was the old paradigm; "de-engineering" the river and letting it run free is the new. By whatever name it is called, the change means improving the environmental quality of a river. In short, this transformation means using a healthy river as an amenity. That is the direction of change, which has been and is continuing to take place.

This change, this transformation, is made possible by several powerful macro-forces, most of which have had no dependence on the good works

The James River as it flows through downtown. *"Sunset Over Richmond"* by Richard Kidd. *Courtesy of Scenic Virginia.*

of local leaders and river advocates. Developments in technology and the economy made water power no longer the most cost-effective source of power and thus pushed industry away from the riverbanks. Simultaneously, leisure time became more abundant along with the resources to indulge in multiple forms of recreation, including time on and along the river. The environmental movement writ-large included rivers as an important element. These forces have provided the foundation for the restoration of the James River in Richmond and many other waterways around the country.

Clearly, however, using a healthy river as an amenity is an oversimplification of what has occurred to this point in the modern period and perhaps is a utopian view of what is possible, or at least likely, in the future. Multiple obstacles, including conflicts among diverse users, stand in the way of the vision coming to full fruition. One might argue that the direction of change has a secure future and is a justifiable source of optimism, but the pace, extent and exact character of that change is uncertain; risks still exist. The seventy-year history described here gives substance to that generalization.

The transformation of this urban river, the James River as it flows through Richmond, Virginia, can be understood in two ways. One is to consider it the story of a special, one-of-a-kind river. After all, the James River has a special place in American history, and the Richmond reach of that river is unusual in the beauty and variety it offers. To treat it solely as the account of a single river undervalues the significance of the story, however. The second way is to consider the story much more broadly as a representative of many rivers nation- and worldwide.

A BRIEF LOOK AT THE JAMES RIVER IN RICHMOND

The James River in Richmond has a claim to uniqueness because of its history and the variety it offers. Historically, the James at the fall line rapids where Richmond now stands was a center of Indian culture. The location was visited by English settlers in 1607 within days of arriving at Jamestown and was the place of an early meeting between Indians and English. Through the development of early industry, the Revolutionary War, the slave trade and the Civil War, Richmond, with the river playing a significant part, was a prime location for the unfolding of American history.

The James River in Richmond also is exceptional because of its diversity. The ten-mile river reach has the makings of:

- abundant wildlife *and* a vibrant downtown
- a sometimes unruly river *and* a historic restored canal
- whitewater experiences *and* flatwater for quiet paddling and motorized craft
- a riverside walk with waterside cafés *and* secluded islands and riverside hiking and biking trails with not a building in sight
- riverside high-rises *and* river access for the public
- important historic sites preserved *and* brand-new architecture
- a tourist destination *and* recreational opportunities for the residents of Richmond.

Although the Richmond riverfront is far from the largest or the most commercially active, few cities in the country can offer a combination this varied, this rich, this attractive. Several decades ago, that rich tapestry was simply potential, blocked by pollution and disinterest. The transformation from potential to a still emerging reality is the story to be told.

Topography provides the basis for the variety on the James River in Richmond; human activity over time has added to it. In the roughly ten

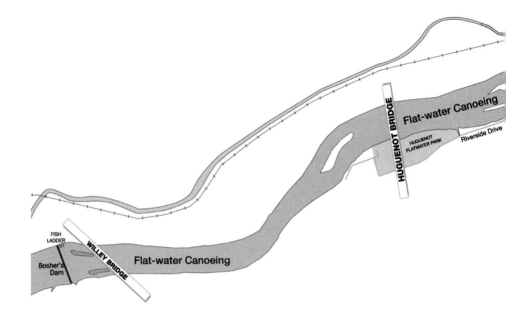

These pages and following: Map of the Richmond reach of the James River. *Courtesy of James River Park System and Sattler Creative.*

miles within the city, the James changes from flat, moving current to foaming whitewater to a slow-moving tidal river. The islands and shoreline range from undisturbed nature to an urban downtown. A variety of historical sites and artifacts are scattered along the way. Parts of the river have been protected from development, other parts highly developed.

Recognizing, preserving and enhancing this variety has been a dominant theme in the transformation of the river. Whether intended or not, this variety has been the pattern of activity over the past seventy years. That is not to say that every decision has been the best or that progress has been achieved as rapidly as it could have been or that the effort has been without obstacles.

The history and diverse features of the James River in Richmond add charm, beauty and potential but also make the transformation more complex and, in many ways, more difficult. Floods have prevented unwanted development in many areas along the river, helping to preserve the urban wilderness that is now highly valued. But these floods also have brought about the construction of a floodwall in other parts that impedes access and blocks views. The rail lines, too, have stopped incursions that

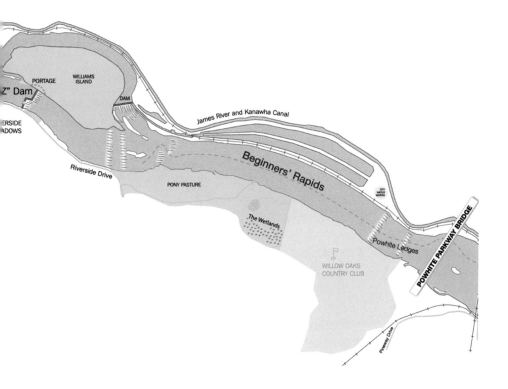

would be destructive of the potential for a park-like future, but they also inhibit the unobstructed access and views that would otherwise be possible. Historical sites along the river, including canals, make it a richer environment for human experience, but historical preservation can be a constraint on what is possible.

As rivers of the world go, the James River in Richmond is a tiny speck, and it is only a small portion of the full 340 miles of the James River, the longest river entirely contained in a single state. Yet this reach has beauty, diversity and charm packed into a short 10 miles that is matched in few places elsewhere. The James in Richmond begins with a white cascading wall of water flowing over Bosher's Dam, originally constructed in 1823 and rebuilt in 1835 (see map segment on page 18). For the next 2 miles or so, the river flows smoothly, with occasional rocks protruding above the surface—except, of course, in high water, when all rocks are covered as the current flows with greater urgency.

The bank on river-left ("river-left" refers to one's left while facing downstream) for much of this section is privately owned undeveloped floodplain woodlands with trees, vines and wildlife relatively undisturbed.

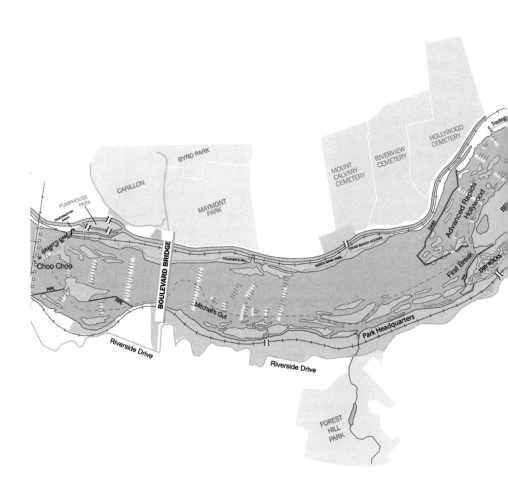

After passing the Huguenot Bridge, river-left sports several homes—all prepared one way or another for flooding. On river-right from Bosher's Dam farther downstream, the bank rises steeply and the slopes are sprinkled with homes that overlook the river, some plainly visible, others camouflaged among the trees. Riverside Drive runs adjacent to the river for most of this stretch and is used as much by pedestrians and cyclists as motorists.

This reach from Bosher's Dam to Williams Island and Williams Dam is a popular sheet of flatwater dotted in the summer, spring and fall with canoes, kayaks, john boats with trolling motors and, increasingly, stand-

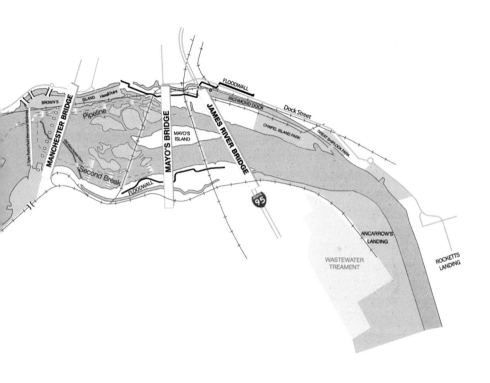

up paddleboards. Although summer is the highpoint of use, it is not abandoned in winter; even on a cold winter day, a paddler or fisherman is likely to be seen. Geese and osprey are abundant, and the occasional bald eagle can be spotted. Public access to this portion of the river is provided by the James River Park System at Huguenot Flatwater, just upstream from Huguenot Bridge.

At Williams Island, ninety-five acres of wilderness owned by the city, the river splits into two channels with a lowhead dam on each side of the island (see map segment on page 19). As the flow passes over the dams, the current picks up speed and the character of the river changes. For

Bosher's Dam marks the beginning of the Richmond segment of the James River. *Photo by author.*

Paddlers frequent this section made accessible by Huguenot Flatwater, part of the James River Park System. *Photo by author.*

the next several miles, the James is a series of riffles and small rapids separated by long pools of slow-moving water.

On river-right, less than a mile downstream of Williams Island Dam (more popularly known as Z Dam), is Pony Pasture Rapids. Pony Pasture, a unit of the James River Park System, has been a summertime magnet for swimmers and rock-hoppers—to the sometimes consternation of nearby residents—since well before the establishment of the park. The Wetlands, a downstream extension of Pony Pasture, is a nature preserve with a maze of walking trails.

Downstream from Pony Pasture and The Wetlands, very few homes are near the river. Rather, the banks and bluffs are the location of other man-made features, including Willow Oaks Country Club Golf Course on river-right with the settling basin and water treatment system for the city, the rail lines and the bed of the James River and Kanawha Canal (sometimes just called Kanawha Canal) on river-left. For the most part, these man-made encroachments on nature do not meet the eye of a paddler and so do not detract from the natural ambience of the river. Two bridges do signal the

Pony Pasture Rapids, seen here from upstream, is one of the most popular sites in the James River Park System. *Photo by author.*

urban location of the river: the Powhite Parkway Bridge (1973) and the iconic arched Belt Line Railroad Bridge (1919).

Past these bridges are a number of islands, several relatively small rapids, including Choo Choo and Mitchell's Gut, and a sewer line that unfortunately was not buried below the riverbed (see map segment on page 20). On river left is the historic Byrd Park Pump House, the Carillon, Maymont. Looking downstream, the city skyline begins to come into view.

The Reedy Creek section of the James River Park on river-right comprises numerous small islands with limited access and a flat floodplain shoreline that runs one and a quarter miles from Boulevard Bridge to Lee Bridge. This section of the park contains the Reedy Creek Visitor Center and Park Headquarters, hiking and biking trails along the river that connect uphill to Forest Hill Park and a canoe/kayak takeout and launch site, which is the primary put-in for running the larger rapids farther downstream.

Not far downstream of Reedy Creek at the western end of Belle Isle, the river makes another change in character. As it flows through the breaks in Hollywood Dam, with historic Hollywood Cemetery visible on river-left, the river's pace again accelerates. For the next mile or more, the "falls" add to the skill level required for paddling. The river is now "downtown," so high-rise buildings dominate the skyline, especially on river-left. The flow passes on the north side of Belle Isle creating the well-known Hollywood Rapid. The river's flow is blocked on the south side of Belle Isle, except in very high water, by the remnants of Belle Isle Dam. The resulting dry riverbed is a dramatic boulder garden.

Belt Line Railroad Bridge is a popular subject for photographers and the site of Choo-Choo Rapid. *Courtesy of Scott Adams.*

The gradual appearance of the city skyline provides a dramatic view for boaters. *Photo by author*.

Rafters are part of the entertainment for sunbathers on the rocks by Hollywood Rapid. *Photo by author*.

The river flow passes under the Lee Bridge, with the underhanging pedestrian walkway that provides access to Belle Isle, then under the newly constructed and much celebrated T. Tyler Potterfield Memorial Bridge atop Brown's Island Dam, past Brown's Island, under Manchester Bridge, through several islands, under Mayo's Bridge and into the slow-moving tidal portion of the James (see map segment on page 21). Multiple channels through largely undisturbed islands provide options for downtown paddlers, although the pipeline series of drops is by far the most used route.

Where the tidal reach begins, the river again changes character. The Interstate 95 Bridge, far more massive than the historic Mayo's Bridge just upstream, seems to announce a change in scale for the river itself. As the fast-moving water joins the more sluggish tidal river, the floodwall, especially on river-right, displays its intrusion on the natural landscape.

Continuing past Mayo's Island and Chapel Island, the river ambles toward the eastern boundary of the city on the north side at Orleans Street. Along the way, sights on river-left include Great Ship Lock, the Soldiers and Sailors monument atop Libby Hill, the mouth of Gillies Creek, Intermediate Terminal and the new development of Rocketts Landing in Henrico County.

The series of rapids along pipeline is the most popular of the downtown routes. *Photo by author.*

Historic Mayo's Bridge, with a rail bridge and I-95 beyond, signals the end of the whitewater trip. *Photo by author.*

The tidal portion of the river has a character quite different from the fast-moving water just upstream. *Photo by author.*

Out of sight on river-right is the wastewater treatment plant. Before the curve on river-right is Ancarrow's Landing, another part of the James River Park System. It is the terminus of the Slave Trail, a popular place for shoreline fishing, a launching point for motorized craft and a busy place during the spring shad run. The south-side city boundary is much farther downstream past the Richmond Terminal, a working commercial dock.

THE LAST SEVENTY YEARS: ELEMENTS OF THE TRANSFORMATION

By the 1950s, water power was declining as the preferred source of industrial power, so industry along the river was in decline or abandoned. The river itself was the sewage treatment facility and otherwise ignored by most. There was virtually no public access to the river, although there were some who fished the upper reaches, partied on the rocks at Pony Pasture or ventured on the water in the seventeen-foot Grumman aluminum canoes that were the standard of the day. But for the most part, the river was an ignored wasteland. Certainly, it was not on the official agenda as a public resource. That began to change.

The transformation of the river did not take place overnight, and for some advocates it has been agonizingly slow. Nor has the change required only a single strategy. The transformation has many elements: building a sewage collection and treatment system (chapter 2), turning back the proposed riverside parkway (3), establishing James River Park (4), restoring and maintaining wildlife habitat (5), facilitating and refereeing the growth of recreational activity along the river (6), proposing multiple plans and ideas that built a vision for the downtown riverfront (7), crusading for the construction of a floodwall (8), collaborating to turn vision into action on the downtown riverfront (9), fighting to preserve scenic vistas (10) and making the river a showcase for history (11). The transformation journey not only represents a shift in the health of the river and the physical features along its length but also presents a new way of thinking about—and valuing—the river. A new river ethos is emerging with attitudes and behaviors that bode well for the future of the river (12). The events and developments described in these chapters do not, however, end the story of the river's ongoing transformation. Opportunities, and threats, lie ahead.

Each of these chapters represents a substory that stands alone, yet each adds to the mosaic that represents the larger urban river story—a story that portrays the conflict, the cooperation and the passion that have been integral to the transformation of the James River in Richmond. While the events portrayed in these chapters overlap in time, it is only fitting to begin with the wastewater collection and treatment story. It is a foundation for the rest.

2

A SEWER NO MORE

CLEANING UP THE RIVER

The rivers of Virginia are the God-given sewers of the State.
—Virginia legislator in 1912

Beside the city's deserted Upper Terminal, Gillies Creek pours raw sewage into the river. At the mouth of the creek, a large circle of green scum widens turgidly. The odor of sewage, and of things long dead, clutches at the nostrils. Here no aquatic life of any sort survives. Our river is a sewer.
—editorial, Richmond News Leader, *1963*

At mid-twentieth century, there was no doubt that the James River in Richmond and miles downstream were severely polluted. A key precondition for the river to reach its modern function was a major cleanup, so perhaps the most significant event in the transformation of the James in Richmond has been the development of a wastewater collection and treatment system. This development has not been fast or without controversy, nor has it been fully concluded. But it has been essential; a river that *is* a sewer does *not* make a park. In 1950, the James River in Richmond was a sewer—or a linear wastewater treatment plant.[1]

Shortly after the Civil War, Richmond began building sewer lines; most led directly to the James River, all without treatment. The only change from 1865 to 1950 was the increased number of sewer lines and the volume of raw sewage deposited in the river. Conditions are graphically described in a 1949 article:

The city of Richmond alone discharges about 13 tons of settleable solids into the James River every day. This generous donation is partly responsible for the fact that 40 miles of river below Richmond have been killed for commercial fishing because fish from this section have a bad odor and a bad taste. The James River is sick from one end to the other, but for 14 miles below Richmond it is entirely dead…except for some very much alive bacteria.

The vivid description continues:

As a recreation stream the James lost its use long ago. Naturally, this foul condition hurt the river for navigation, both for small craft and commercial vessels. Some 1,200 yacht owners in Richmond make it a point to keep their boats off the James, not only because they know that pollution will attack the painted bottoms, but also because they do not like having their sense of decency insulted by watching the sewage of 225,000 human beings float by on the waters of what should be the beautiful James.[2]

A SLOW BEGINNING

From time to time in the first half of the twentieth century, there was discussion of the need to clean up the river, but little or nothing was done. Efforts to impose controls were turned aside by lobbyists who considered them not cost-effective. The most famous and oft-quoted statement was that of a Virginia legislator who stated, "The rivers of Virginia are the God-given sewers of the State."[3] The words of most of those opposed to cleanup action were not nearly so direct or quotable, but the effect was the same—no action.

Even after the Virginia General Assembly enacted the State Water Control Law in 1946, which established the State Water Control Board (SWCB), localities, including the City of Richmond, resisted its requirements. The mission of the board was to "protect existing water quality, to reduce and prevent water pollution, and to restore and maintain state waters to a quality that would protect human health and aquatic life."[4] Shortly after it was formed, the SWCB told the City of Richmond that it must apply for a permit to dump sewage into the river. Richmond made application and it was granted, giving the city permission to continue to dump raw sewage into

Sewage flows in the James River. A graphic illustration of decades of practice.
Courtesy of Richmond Times-Dispatch Collection, The Valentine.

the river, at least for a time. The 1949 board told the city that it "shall" work to develop a program of pollution reduction. Richmond responded that it would take six years to develop a treatment facility; the board said that was too long. This was the beginning of a series of contests between the city and the state board. The city's consistent position was "we would like to do it, but we don't have the money and we need more time," while the board insisted that it must comply in a timely manner.

A clear example of the city's argument was made in a *Richmond Times-Dispatch* editorial in 1949 in response to an early consulting engineers report that $13 million would be required to build a sewage disposal system: "We endorse and believe in the antipollution program, but the polluting has been going on for generations, even centuries, and we do not feel that Richmond's present financial condition justifies the expenditure of $13 million for one single capital project over the next four years."[5]

The city's early delay action was assisted by a Richmond delegate to the General Assembly, W. Griffith Purcell, who introduced a bill in 1952 calling for a two-year delay in any order the State Water Control Board might issue to force Richmond to construct a sewage disposal facility. Purcell argued that the city had more pressing needs, that cities downriver of Richmond got their water from sources other than the James River and that pollution of the James comes mostly from Lynchburg and industrial plants upstream of Richmond.

Delegate Purcell's argument was supported by former mayor J. Fulmer Bright, who said he would fight any plan to build a sewage treatment plant. He said the city had to decide whether it would or would not obey the state water control act.[6] The *Richmond News Leader* argued for support but delay. The city should prepare plans and have them ready but delay until the financial situation improved. In the meantime, "ending Richmond's traffic congestion is still more urgent and hence should have a priority claim."[7]

DESIGNING AND BUILDING A WASTEWATER TREATMENT SYSTEM

Despite "support but delay" objections, Richmond was moving forward with a response to the prodding by the State Water Control Board. Several consulting engineers had conducted studies and prepared cost estimates, and in 1950, the city hired Greeley and Hansen, a Chicago firm, to develop a more detailed plan.

As further indication of movement in 1952, members of city council and others set out on a "sewage safari" down the river below Richmond to see river conditions firsthand. The public works director, Jack Gould, expressed his regret that recent rains had raised the water level and the safari would not see the river at its most offensive.[8]

By the end of 1952, the decision had been made to proceed with wastewater collection and treatment, and in January 1953, the lead engineer for Greeley and Hansen arrived in Richmond to begin the detailed design work. The two key elements in the system were (1) the interceptor sewer along the river to collect sewage from the pipes that were dumping directly into the river and (2) the treatment plant itself.

In October 1958, the first phase of the sewage treatment plant was dedicated and put into operation. While a significant accomplishment, the plant provided only primary treatment and only a portion of the interceptors were complete and sending waste to the plant. Additional interceptor connections were scheduled to occur over the next several years.

Reminiscent of the sewage safari made by city council eleven years earlier, the *Richmond News Leader* staff made a power boat inspection tour of the river below Richmond in 1963. Its interim "progress report" did not indicate much progress:

> The mixed impression leaves us discouraged and hopeful all at once. The river's surface is a dull and greasy green. Its currents move sluggishly; by the shoreline, bubbles of methane gas, caused by rotting materials on the river floor, pop and twinkle on the surface. Tangled underbrush at the water's edge holds captive dead and dying fish. Overhead turkey buzzards wheel in lazy circles spiraling down upon their prey. A large catfish struggles to the surface, gasping. It dies. Beside the city's deserted Upper Terminal, Gillies Creek pours raw sewage into the river. At the mouth of the creek, a large circle of green scum widens turgidly. The odor of sewage, and of things long dead, clutches at the nostrils. Here no aquatic life of any sort survives. Our river is a sewer.[9]

Despite the negative assessment, the editorial concludes with an eloquent statement of the potential beauty and recreational value of the river, an indication that attitudes toward the river might be changing.

Although the cleanliness grade in 1963 was poor, work was continuing to improve wastewater treatment. By that time, only primary treatment of wastewater had been accomplished, but upgrading to secondary treatment was well underway. The secondary treatment plant was scheduled to go online in 1972 when the flood from Hurricane Agnes hit, delaying the opening. It did open in 1973. (Primary treatment involves the separation and removal of insoluble matter and solids with screens and sedimentation; secondary treatment uses biological processes to degrade organic material.) Upgrading the treatment plant was and is an ongoing process, continuing to the present.

In the 1970s, federal policy change was coming online to provide both carrots and sticks for river cleanup in Richmond and around the nation. The National Environmental Policy Act became law in 1970, followed by the Federal Water Pollution Control Act amendments (amending the Federal Water Pollution Control Act of 1948). These 1972 amendments, known as the Clean Water Act (CWA), required setting water quality standards and obtaining a permit for point source discharge into streams and also provided significant funds for wastewater treatment upgrades. It worked through states, imposing requirements states could enforce, but the federal government reserved the power to intervene if state actions did not meet standards.

While the wastewater treatment facility was undergoing improvements and sanitary sewers were no longer dumping directly into the river, an additional reason for the still polluted river was that nothing had yet been done about combined sewer overflows. That was to be the next big problem for the city and the next battle between the city and the State Water Control Board with active participation by the U.S. Environmental Protection Agency, individual citizens and several environmental advocacy organizations. Resolving the combined sewer overflow problem has taken decades to play out and is still ongoing.

COMBINED SEWER OVERFLOWS

When the sewers were built in the older downtown section of the city in the late 1800s and early 1900s, the state of the art was to run sewage and stormwater into the same pipe. After all, it was all going to the river anyway. The last combined sewer lines were constructed in the 1940s; at the beginning of the twenty-first century, approximately 30 percent of Richmond was served by a combined sanitary-stormwater system.[10] The interceptor pipes that were installed to carry wastewater to the treatment facility worked well in dry weather, but not during rain events. The conveyance pipes and the wastewater treatment plant could not handle the combined sewage and rain runoff, so there were overflows that spilled untreated into the river. This became known as the combined sewer overflow (CSO) problem.

Resolving the combined sewer overflow issue along with upgrading the treatment plant to secondary and tertiary levels have been the major challenges since the 1970s. The story line has been much the same as the fifties and sixties period—the State Water Control Board, and increasingly the federal government, prodding and the city insisting that it could only go so fast and that the state and federal levels should contribute more to the cost. The locus of front-line responsibility within the city changed in 1969 when wastewater responsibility was shifted from the department of public works to the department of public utilities.[11]

During the early period of river cleanup, the most vocal spokesman for action was Newton H. Ancarrow. He was a regular visitor and speaker at city council and State Water Control Board meetings and is reported to have brought dirty James River water to meetings of council to illustrate his claims the river was foul. (How vile the water was that he displayed and what it contained sometimes differ in the telling.) Ancarrow was successful in bringing the pollution problem on the James to the attention of the Environmental Protection Agency even before the passage of the Clean Water Act.[12] Further, he brought suit against the city, claiming that the pollution destroyed his ability to use the river for his marina and boat building business.[13] His suit failed, and the city later condemned his riverside property to build the sewage treatment plant. (Some say a dose of retribution was involved in the condemnation decision.)

The city and its consultants considered two general approaches to the CSO problem, neither cheap: (1) separation of the combined system or (2) construction of large holding basins to contain overflows until the excess volume could be treated. The option of separating sewers and stormwater lines was considered too costly and disruptive almost from the beginning, but the holding basin option became an integral and continuing feature of the CSO plans and action. Early in the planning process, a consultant's report recommended using the James River and Kanawha Canal Basin between 17th and 27th Streets, also known as Richmond Dock, as a holding basin for combined sewer overflow in the Shockoe area. This site, while considerably less expensive than other localities, did not prove to be a popular choice.

Richmond Dock, the final segment of the James River and Kanawha Canal, almost became the combined sewer overflow retention basin. Protests from multiple quarters prevented that from happening. *Photo by author.*

Despite growing objections, the use of the canal basin did receive approval from several bodies, including the Richmond Planning Commission, Richmond City Council and the Richmond Regional Planning District Commission. The director of the Richmond Department of Public Utilities argued that the odor generated would be slight and that the concrete walls and bottom that the conversion to a retention basin would entail would improve the canal. Opponents of using the site as a retention basin did not buy these arguments. The Church Hill Civic Association, conservation and historic preservation groups, private citizens and some businesses raised their voices against the plan. Their concerns included destruction of an important historic structure, fear of odor and other negative effects on the area and even an expression of hope that the canal system, including Richmond Dock, could someday be restored as a historic landmark and recreational facility. The opposition carried the day, and Mayor Thomas J. Bliley Jr. announced plans to abandon the canal site and consider other means for storing the overflow water and sewage. The scheduled State Water Control Board public hearing to consider the canal basin was postponed indefinitely, and the board gave Richmond a twelve-month delay to develop another plan.

In 1972, a study was released that confirmed that containing the overflow in the Shockoe Creek (now a pipe, no longer a creek) area would have the greatest beneficial impact on the river and that the construction of a retention basin would be the most effective approach. A follow-up report in 1973 suggested building this retention facility on the south side of the river closer to the treatment plant. Ultimately, however, the north side of the river on Chapel Island adjacent to Richmond Dock was chosen. Despite earlier conclusions that separating sanitary from stormwater drains was too expensive, there were some, or at least one, who argued that separation would be the most cost-effective approach in the long run. Newton Ancarrow claimed that the only effective way to achieve the "ultimate goal" of a clean river is to separate the two. "I am told," Ancarrow stated, "it would cost upwards of $300,000,000 to separate the sewers in Richmond. So be it." Separation is the only cost-effective option and "the final goal of a clean viable river" cannot be achieved "with the present mishmash of reactions and halfway measures."[14] Ancarrow's advice was not heeded.

Shockoe Retention Basin interior, a "closed container" for the first flush in a rainstorm. *Courtesy of Mariane Jorgenson, Richmond Department of Public Utilities.*

The Shockoe Retention Basin, completed in 1983 at a cost of $43 million,[15] was designed to retain the "first flush" of combined sanitary and storm sewage during a rain event that overwhelmed the lines to the wastewater plant. The first runoff in a storm is the most contaminated, so even if the entire volume is not captured, the worst of it is. After the wet weather subsides, the retained sewage and stormwater is piped to the wastewater treatment plant on the south side of the river. The Shockoe Basin is 16 feet deep, covers 150,000 square feet in area and, after a major rain event, can be emptied in about two days.

(An expensive inconvenience arose in 2008, twenty-five years after the Shockoe basin opening. Although the retention basin was equipped with an aeration system to keep solids in suspension until the "sewage stew" is emptied, a ten-foot layer of solid sewage had accumulated over the years. No provisions had been made for clean out in the original design, so the city, at a cost of $12 million, had to cut a door, build a ramp and use backhoes to remove the solidified sewage.)[16]

TOWARD A COMPREHENSIVE CSO PLAN

Over the next several years, the "arm wrestling" between the City of Richmond and the State Water Control Board continued with studies begun and put on hold and begun again. Although the completion of the Shockoe Retention Basin and the continued improvement of the wastewater treatment plant were significant accomplishments, for nearly a decade, the city had devoted little sustained attention to achieving a comprehensive CSO solution even though there were calls for action by external groups. At least two environmental organizations, the Lower James River Association (later to become the James River Association) and the Chesapeake Bay Foundation, asked the State Water Control Board to require the city to develop a plan to address the continuing CSO problem.[17] As downriver recipients of Richmond's waste, these organizations had an understandable interest in the issue. In 1985, the development of a CSO plan again became an important feature of river cleanup when the State Water Control Board ordered the resumption of planning for a comprehensive solution.

Although recognizing that CSO discharges were a problem that must be addressed, city officials were still reluctant to move quickly. Again, the primary issue was money. The city maintained throughout that it could not bear the cost of cleaning up the river alone, that state and federal assistance were required. Imposing the full cost on residents through hiked utility rates could go only so far, especially, as the city argued, since a large portion of residents were low income. A second source of reluctance was uncertainty about the effectiveness of various approaches.

Even A. Howe Todd, assistant city manager and frequent river advocate, was quoted as saying the millions already spent on the Shockoe Retention Basin haven't yet proven to be worth the cost. "Let's be sure what we're doing before we spend millions. Millions of dollars are hard to come by." Arguing that the pollution that occurs is mainly in the tidal area and that the river has great diluting power, Todd said, "Nobody has yet proven that the dumping of that sewage into the river at the time of a heavy rain is that big of a problem because of the diluting effect of the river."[18]

Despite the concerns about cost and uncertainty about the best solution, the city did agree to meet the new wastewater treatment plan requirements and the demand to complete the comprehensive CSO study. These requirements were part of the Virginia Pollutant Discharge Elimination System (VPDES) permit and Consent Decree issued in 1985 and extended to 1990. An agreement was reached, signed and ultimately ratified by the circuit court over the objections of environmental and recreational organizations, including the Environmental Defense Fund, the Chesapeake Bay Foundation and the Lower James River Association. The primary offending features from their perspective were that the agreement allowed too much flexibility by making some deadlines contingent on funding and extended the date by which permit limits must be met to 1990 rather than 1988, as required by federal law. And, as Patricia A. "Patti" Jackson, former executive director of the James River Association, emphasized in a recent conversation, more than two dozen CSOs had continued to contaminate the river.

THE COMBINED SEWER OVERFLOW STUDY OF 1988

The Combined Sewer Overflow Study Final Report of 1988, prepared by the engineering firm of Greeley and Hansen, became the centerpiece of the response to the CSO problem.[19] This cost-effectiveness study examined thirty preliminary alternatives and, in more detail, eight final alternatives.

The selected plan would be constructed in three phases. The first phase was to improve the wastewater treatment plant to make it possible to empty the Shockoe Retention Basin in two days. This project was already underway and would be part of any plan developed. New commitments began with the second phase, which would involve the construction of conduits on the north and south sides of the river to convey CSO out of the James River Park area and also included construction of disinfection facilities. The third phase would involve the construction of conveyance conduits for the remaining CSOs and construction of treatment for all CSOs, including swirl concentrators and retention basins.[20]

Again, not unexpectedly, the first concern on the part of the city was cost. The report projected capital costs for phase 1 at $73 million; for phase 2 at $40 million; and for phase 3 at $181 million, for a total of $294 million.[21] The report itself recognized that the city alone could not pay the total cost and included a "grant funding strategy" section.

A second issue, this raised by the Falls of the James Scenic River Advisory Committee, was the impact on the river and riverbanks. The committee,[22] although in favor of the objectives of the plan, noted that conduits were proposed to be placed at the edge of the river and that portions of the conduits would be visible along the banks. This location the committee argued "would destroy the beauty and integrity of the natural riverbanks which are the most important features of the parks."[23]

A third objection to the 1988 CSO plan came from the downriver environmental groups, which argued that much of the pollution from CSOs, although largely diverted from sensitive areas in Richmond, still entered the river. Downstream reaches of the river and the bay were not protected.

The wrangling over CSO plans continued for a number of years with environmental group objections, plan revisions, EPA and SWCB disapprovals extensions and permit approvals. The city's plan was to proceed as federal grant funds allowed, while environmental groups argued that it was illegal to make action contingent on grant funding.[24] At first, the EPA disapproved the plan,[25] and a yearlong stalemate followed. The agency then approved the plan[26] that still was assailed by the environmental groups.[27] The SWCB delayed its decision by a month[28] and then approved the plan in January 1992.[29]

A significant political development during this period was the organization of a national CSO partnership led by Richmond mayor Geline Williams. Richmond joined with cities across the country also struggling with the CSO problem to seek solutions and to lobby the federal government for funding.

While the struggle to develop and fund a CSO plan was occurring, the city completed and held the grand opening of a major upgrade to the wastewater

Left: Combined sewer overflow (CSO) pipeline installation in the restored canal bed. The collaboration made both projects more cost-effective. *Courtesy of Venture Richmond*.

Below: Inside the Hampton-McCloy Retention Tunnel, a massive underground structure that captures the combined sewage-stormwater runoff and then pumps it to the wastewater treatment plant. *Courtesy of Robert Stone, Richmond Department of Public Utilities*.

treatment plant. This multimillion-gallon plant included a "tertiary effluent filtering system" with a capacity of ninety million gallons per day. The water returned to the river was said by the department of public utilities to be cleaner than that collected upstream.[30]

The physical improvements to the CSO system in the 1990s and early 2000s continued with three important elements. One was the installation of a conveyance pipe under the river along the south bank. The second project, the north-side conveyance, was even more ambitious. It included a pipeline with a diameter of ninety-six inches from the Lee Bridge to the Shockoe retention basin on Chapel Island.[31] The unique feature of this project was combining the CSO pipeline installation with restoration of the Haxall Canal and development of Canal Walk (see chapter 9). The pipeline and restored canal were constructed simultaneously, with the pipeline installed beneath the bed of the canal. Joint financing with federal and local funds made each outcome more affordable. This serendipity—or brilliant planning and admirable cooperation—has been a signature achievement of downtown riverfront transformation.

The third accomplishment of the period was the construction of a 7.2-million-gallon retention basin tunnel on the north side near the Powhite Parkway Bridge. The Hampton-McCloy Retention Tunnel completed in 2003 at a cost of approximately $122 million is six thousand feet long and sixteen feet in diameter, lies one hundred feet below riverbed level and extends under Maymont Park.[32] As with the Shockoe Retention Basin, the first flush overflow is stored and then pumped to the Wastewater Treatment Plant.

In 1999, the city received the National Combined Sewer Overflow Control Program Excellence Award from the Environmental Protection Agency[33] and also the Friends of the James Award from the James River Association, one of the organizations that had pressed so hard in the early nineties for a solution to the CSO problem.

BEYOND THE AWARDS

Despite the awards for addressing the combined sewer overflow problem, the job of cleaning up the river was far from complete. Subsequent phases have continued to upgrade the wastewater treatment plant and the facilities for handling and treating combined sewer overflows. Nevertheless, the goal of zero discharge events is still some years and millions of dollars away.

Discharges from some CSO outfalls occur on average less than once per year; others discharge more frequently. (The department of public utilities provides information about overflow events on its website and through an email service.) The Gillies Creek watershed, which empties into the tidewater section of the James, the largest remaining CSO problem area, will be especially costly. To get Gillies Creek to fewer than four overflows per year, according to former department of public utilities (DPU) director Robert Steidel, is at least a decade away, and the capital cost is likely to exceed $60 million without state or federal help. The city portion of Gillies Creek was paved with concrete in 1974 and, except during rain events, carries very little water. Cleanup expectations for the area have been raised since the creek crosses under the recently completed Virginia Capital Trail and through an area marked for riverfront recreational development in the 2012 Riverfront Plan (see chapter 9). Turning Gillies Creek into a greenway is in the discussion stage.

Sanitary wastewater and combined sewers are not the only issues regarding river pollution. Stormwater that does not flow into the combined sewer system and does not mix with sewage but nevertheless may flow without treatment to the river is also a source of pollution and a focus of local concern as well as federal and state laws requiring local action. Heavy precipitation picks up contaminants such as oil, pesticides, fertilizer and animal waste and, unless dealt with appropriately, ends up in streams and the river. Approximately two-thirds of the city is served by a separate storm sewer system and a complex of ditches, pipes, ponds, streams, catch basins, floodplains, curbs, gutters, storm drains and wetlands, all of which require attention and dollars. The purpose of the system is to prevent flooding, enable the recharging of groundwater, prevent erosion and reduce pollutants entering waterways.

In 2007, responsibility for stormwater management in the City of Richmond was transferred from the department of public works to the department of public utilities, and in 2009. Richmond City Council established a stormwater utility within DPU. The utility structure enables the collection of fees from both commercial and residential properties in proportion to their contribution to stormwater runoff. The fees in turn support a multipronged effort to manage stormwater and reduce pollutants entering waterways.

Federal policy again added new marching orders for the City of Richmond (and other cities, states, businesses and farmers) when the Environmental Protection Agency announced the Chesapeake Bay TMDL (total maximum daily load, aka "pollution diet") at the end of 2010. This TMDL, the largest ever promulgated, set maximum amounts of pollutants (in this case nitrogen, phosphorous and sediment) that could enter the bay and still meet federal water quality standards. The acceptable load was divided among the multiple jurisdictions in the Chesapeake Bay watershed. The Bay TMDL is now a significant driver of pollution abatement efforts in Richmond.

Reducing stormwater pollution is a significant requirement of the TMDL, and stream restoration makes up a piece of Richmond's strategy to achieve its pollution reduction targets. The city, through DPU, is working on or planning four restorations: Goode's Creek, Pocosham Creek, Reedy Creek and Rattlesnake Creek. The city stormwater utility would pay for 50 percent of each restoration, and matching grants from the Stormwater Assistance Fund administered by the Virginia Department of Environmental Quality would cover the remainder. Combined costs of the four projects is estimated at $5.75 million.[34]

Stream restoration would seem to be an environmentally sound and universally applauded undertaking, but not in every case, as DPU discovered. The proposed Reedy Creek restoration raised a storm of protest from residents in the area—with the concern that the project was environmentally unsound. The central argument of the objecting Reedy Creek Coalition is that the design of the DPU proposal is risky since the restoration only addresses a downstream portion of the creek, leaving unchanged the upstream portion, which contains a one-plus-mile concrete channel. Disturbing the downstream without first restoring the upstream, the coalition asserts, risks irreparable damage should a major storm send a surge down the concrete channel and pour through the restored area, especially before the new vegetation matures. The Reedy Creek Coalition brought its case to Richmond City Council on November 14, 2016, and the council voted 8–1 to reject the matching grant money from department of environmental quality and 9–0 to delay the project while other alternatives are considered.[35] Passion and conflict once again became part of the river transformation story, but it is important to note that the possibility for cooperation was preserved.

PERSPECTIVE

Great progress has been made in converting the James River in Richmond from the sewer it was when a 1949 article asserted that "14 miles below Richmond is entirely dead," or when the "sewage safari" was taken in

1952, or the staff of the *Richmond News Leader* wrote in 1963 that "our river is a sewer," or when Newton Ancarrow carried "jars of murky James River water to the city council and invited members to take a drink."[36] A striking symbol of this progress occurred in the summer of 2014, when the River City Magnolias, a synchronized swimming group, performed at Ancarrow's Landing,[37] the site where, at one time, pollution reportedly could strip paint from a boat.

With the successful transformation of the river that the cleanup has made possible, one might wonder about the foot-dragging on the part of the city that characterized the pollution issue in earlier days. Could more aggressive action have brought about success much earlier? Were shortcuts taken that still plague the city today? Given the complexity of efforts to resolve the CSO issue, not yet fully successful, one might speculate about the wisdom of Newton Ancarrow's argument that enduring the cost and disruption of separating storm and sanitary sewers would be cost-effective in the long run. Perhaps he was right. Of course, at any given moment there are competing uses for scarce resources, and all good causes cannot be funded.

While it now is appropriate to celebrate progress in cleaning up the river and to recognize that many other improvements along the river would not have been possible without it, it also should be realized that the job is not done. Even with the great strides in Richmond, occasional "CSO events" still occur and officials still recommend not swimming in the river during "first flush" immediately after a rainstorm.

There always will be more to do. Federal and state agencies continue to provide requirements, guidelines and sometimes money, but much of the responsibility lies at the local level. City government is on the front lines with the department of public utilities as a primary agent. Future possibilities for the city include a long list: returning Gillies Creek to a natural streambed and greenway with no CSO spillage, replacing the Shockoe Retention Basin with a tunnel similar to Hampton-McCloy and thereby converting Chapel Island to recreational use, retrofitting public streets with vegetative buffers, restoring tributary streams, increasing best stormwater practices by businesses and residents, incorporating more city property into the James River Park conservation easement and continuing to be among the nation's leaders in state-of-the-art wastewater treatment facilities.

Obviously, the water quality in the river as it runs through Richmond is dependent on more than the action taken in the city. What happens upstream is critical. Even toxins in the river downstream can have upstream effects, as the catastrophic Kepone incident in Hopewell demonstrated. And recent oil, chemical and coal ash spills in other locations and other rivers are a reminder that risk still exists.[38] So, the matter of water quality still requires work by multiple parties—the city, state and federal governments, as well as farmers, pet owners, businesses and residents.

∞

While the early stages of cleaning up the river were lurching along, the Richmond Metropolitan Authority announced a proposed expressway along the south bank that would have altered the character of the river permanently.

THE EXPRESSWAY NOT BUILT

AN EMERGING APPRECIATION OF THE RIVER

But for the river to survive in all its beauty and as the natural gift we must leave to future generations, the public is obligated to take care in how the river is used.
—*David D. Ryan,* The Falls of the James, *1975*

Once destroyed it can never be replaced!
—*Richmond Scenic James Council, 1970*

On Monday, October 24, 1966, the headline in the *Richmond News Leader* read, "Expressway Opens Recreation Vista." The emphasis in the headline and lead picture was the Riverside Parkway, but the full story was the public unveiling of the Metropolitan Expressway System, which also included Downtown Expressway, the Beltline Expressway (later designated I-195) and the Powhite Bridge and Parkway. The portion planned by the Richmond Metropolitan Authority (RMA) that inspired the "Recreation Vista" headline, Riverside Parkway, turned out to be a section that was never built. The proposed Riverside Parkway, if carried out, would have had a significant impact on the river—a catastrophic one according to opponents. The parkway would have created a four-lane limited-access highway along the south bank of the James River roughly from the Huguenot Bridge to the current Powhite Bridge. The parkway would have extended into the river for a considerable stretch from Pony Pasture west, requiring the removal of part of Williams Island to replace the channel filled by the Parkway. It would have had an elevation of up to eleven feet above the current Riverside Drive. Additionally, the plan was to extend the Riverside Parkway, at a later date when demand warranted, on the north side of the river to Parham Road in Henrico County.[39]

IMMEDIATE OPPOSITION

Concern about the proposed Riverside Parkway began to stir even before the public announcement of routes and plans. Seven months prior to plan release, J.J. Jewett suggested "developing an organization for the purpose of preventing, if possible, the construction of a road located so as to destroy the substantial esthetic and real estate values now existing in Chesterfield County between the Belt line and Huguenot Bridge."[40] Additionally, John W. Pearsall submitted his resignation from the RMA Board because of the potential for conflict of interest since he owned property that would be affected by forthcoming plans.[41]

Within days after public announcement of expressway plans, John W. "Jack" Pearsall III sent a letter to neighbors in the Riverside Drive area inviting them to a meeting at the Pearsall home to discuss an alternative route his father had already proposed or full opposition to the Riverside Parkway plan.[42] (Jack's father, John W. Pearsall, was out of the country at the time.) Sixty-two people attended the meeting and formed the Stratford Hills Citizens' Committee with an executive committee of William H. Emory Jr., Rives Fleming Jr., Charles A. Hotchkiss, Millard Binswanger Sr. and John W. Pearsall III.[43]

The committee developed a multipage "Statement of Position" detailing arguments against the Riverside Parkway[44] and requested a meeting with RMA officials to discuss their concerns. In that meeting, a member of the delegation, G. Cameron Budd, challenged the RMA cost figures as well as its construction plan. As Jack Pearsall described a portion of that meeting: "I thought the breakpoint came when Mr. Budd asked what size boulders they would use to line the river and opined that those would wash away in the first flood."[45]

Led by the Citizens' Committee with others joining in, multiple meetings were held, TV coverage arranged, public pressure encouraged and letters sent to the RMA and Richmond City Council. The arguments made, sometimes in great detail, can be summarized in four points: (1) The Riverside Parkway leg of the proposed expressway system would not be cost-effective. The traffic this tollway would attract would not be adequate to justify the road and likely not produce enough revenue to cover operating costs, much less pay off bonds. (2) Alternate routes would be more sensible and effective. The north side of the river along the general route of the canal and rail line was one often mentioned alternative. (3) If the parkway is routed on the south side as planned, it should be modified considerably to make it less destructive and obtrusive. (4) The parkway as planned would destroy one of the most beautiful and accessible sections of the river.

Immediate satisfaction did not seem to be forthcoming. In a letter to the RMA, John W. Pearsall (who had returned from his overseas trip) wrote, "It was with considerable surprise that I read in the paper last evening that the Authority may establish final routes for the proposed thirteen mile Richmond Express System at a meeting called for this afternoon, when it is my distinct impression that the Authority has yet to explore fully certain questions raised as to the Riverside Parkway segment."[46] Nevertheless, Charles A. Taylor, RMA chairman, released a statement the following day: "The Richmond Metropolitan Authority approved at a special meeting of the board yesterday afternoon the routes for the Metropolitan Expressway System. The routes selected are substantially as recommended by the consulting engineers." Several alternate routes for Riverside Parkway were considered, according to the statement, but the original "approved route is the only one which will provide sufficient traffic and revenues."[47]

The RMA statement did conclude by saying "we are always open to suggestions and will continue to give thorough consideration to constructive ideas."[48] Many suggestions, and objections, did follow, and letters peppered the RMA over the next several months. The RMA responded with some design modifications but no significant change. As the *Richmond News Leader* reported in September 1967: "Any alterations on the proposed Riverside Parkway will be relatively minor and will not include major shifting of the route, a spokesman for the Richmond Metropolitan Authority said today."[49]

So, despite the opposition that continued in multiple forms over the next two years, it appeared that plans for the Riverside Parkway were alive and well as 1969 came to a close.

A NEW ORGANIZATION AND
A NEW ROUND OF OPPOSITION

One seemingly insignificant occurrence during this early period of opposition, a Girl Scout hike to say farewell to the river, turned out to be a harbinger of events to come. The organizer, Louise Burke, invited a reporter to join the hike, and his brief article about the hike in the *Richmond Times-Dispatch*[50] found its way into the files of Associated Press reporter Ken Ringle, who, in the early summer of 1970, called her with the suggestion that she start a group to fight the expressway. He also brought her attention to the fact that the Virginia General Assembly had passed the Scenic Rivers Act, which might aid the cause.[51]

Although skeptical about trying to form a group or get involved, Burke did give the idea thought. She packed some environmentally oriented books to take on vacation. One of the books was about an environmental battle in Los Angeles, led by "a plump middle-aged mother of two," she said with a smile, "so I thought, well, I qualify." When she returned from vacation, the thought of organizing opposition was reinforced with a phone call from the city's assistant director of planning, Jim Park, who told her that the RMA was getting ready to float bonds for the whole system, including the Riverside component. So, he told her, "If you have any thought of stopping it, now is the time."[52]

Louise called Mable Young because of a letter to the editor Mrs. Young had written objecting to the Riverside Parkway. Louise also talked with Mrs. Young's husband, Dr. R.B. Young, who suggested in turn that she call John Pearsall, who had been fighting the parkway since its public announcement (if not before). Other phone calls followed, resulting in the addition of Ann and Jack Andrews and Doris and John Hurst to a small group meeting on John and Laila Pearsall's porch overlooking the James River in July 1970.

The Pearsall porch meeting resulted in the decision to organize the Richmond Scenic James Council ("council" to indicate an organization of organizations as well as individuals). Louise was selected as chair, succeeded a short time later by Dr. Young. A period of organization building followed with other individuals and organizations contacted to expand membership. Letters were written to a number of organizations and individuals inviting them to join in the effort and to meet the following month. Among the groups represented at the meeting were Coastal Canoeists, Trout Unlimited, Richmond Ornithological Society, Junior League, the Garden Club of Virginia and the League of Women Voters. Richard D. Obenshain was brought on as the attorney for the group; the original group plus Robert Hicks of Trout Unlimited formed the board of directors.[53]

Associated Press reporter Ken Ringle had not only helped start a process that led to the formation of the Richmond Scenic James Council but also captured a seeming contradiction that was playing out along the James in Richmond: the destruction of a portion of the river with the expressway and the protection of another segment with the opening of the James River Park.

> *RICHMOND (AP)—The city of Richmond, in one of those moves that bring environmentalists close to tears, is spending money simultaneously to both celebrate and molest the city's longtime ecological poor relation, the James River.*
>
> *At one end of Richmond preparations are under way for the opening next month of the $720,000 first phase of the $6 million James River Park an ambitious and incredibly beautiful design threading 2,000 acres of tree-arched, wildflowered riverbank with trails and rustic bridges and giving the Richmond public its first legal access to the James within memory.*
>
> *At the other end of the city, in the newly annexed portion to the west, the city-backed Richmond Metropolitan Authority is preparing to run a four-lane expressway for 2½ miles down the south bank, ripping up trees, paving over riverbank and shadowing the boulder-stream rapids with nylon-stilted forays in the riverbed itself....*
>
> *Yet strangely, the James has been largely ignored for most of this century by Richmond, and only recently in a quest of a civic identity beyond its Confederate monuments, has the city focused attention on the natural asset which was there all along.*[54]

The article from which the preceding excerpt was taken, it was pointed out by those opposing the Riverside Parkway, was carried in a number of state papers, but not in Richmond.[55] The inference drawn, of course, was that the local newspapers were part of the power structure that supported the expressway and did not want to lend support or credibility to the opposition.

DISCOVER THE JAMES ON COLUMBUS DAY

A major event in building the organizational strength of the Richmond Scenic James Council and stimulating further opposition to the Riverside

Parkway plan was "'Discover' the James River on Columbus Day" held on October 11, 1970. Building on Sierra Club strategy that people understand an issue best when they see it, the public was invited to visit the riverbanks and see where the Parkway would be located and what it would be like. Ringle's contradiction, or irony, again surfaced. October 11 also turned out to be the day the new James River Park was dedicated a few miles downriver.

On the day of the event, the weather was beautiful. Visitors parked in the River Road Shopping Center parking lot and were shuttled to the riverbank in station wagons marked with ribbons on their antennas. The "tour" started at what later became Huguenot Woods Park and still later Huguenot Flatwater but was private property at the time. Scouts in dress uniform served as guides, and biologists from the University of Richmond and Virginia Commonwealth University gave nature walks. When the visitors emerged at the beach at Rattlesnake Creek, Coastal Canoeists provided canoe rides over to Williams Island to those who wished to go. The water level was low, so after a time on Williams Island, people could walk over Williams Dam (aka Z Dam) back to the south bank of the river and Riverside Drive. (Today, walking to and from Williams Island even in low water is no longer possible because of the fish passage notch cut in the dam in 1993; see chapter 5.)

On the south bank below Williams Dam, the organizers had displays that dramatized the impact the Riverside Parkway would have. The plans for the parkway were displayed, but even more dramatic—and to make sure no one missed the message—there were drawings showing big walls of riprap, red ribbons strung in the trees at the level the parkway would reach and a marker on the dam 185 feet out into the river where the parkway would extend. The tour included the area that is now Pony Pasture Rapids, part of the James River Park System, but at the time was private property.

The "Discover Day" was considered by the organizers to be a galvanizing event. The beauty of the location was apparent, and the extent of the impact of the proposed parkway was made visual. One influential garden club member is reported to have said as she walked through the Pony Pasture area: "For anyone to even think of destroying this magnificent place is absolutely iniquitous." The Richmond Scenic James Council made a lot of friends on Discover Day. "Not as many as we would have liked, but enough to really help us. We even began getting money from anonymous benefactors."[56]

One of the visitors on Discover Day was Alan Kiepper, Richmond's city manager. He was one of the few city officials in attendance, apparently because of the overlapping dedication of the James River Park. Several members of the Richmond Scenic James Council took the opportunity to provide information

The shoreline as it might have looked in 1970 when the protests against the proposed riverside expressway took place. *Courtesy of Scott Weaver.*

A segment of the river that would have been covered by fill, concrete and asphalt had protest and economics not altered expressway plans. *Photo by author.*

to Kiepper above and beyond that conveyed in the displays. Transportation specialist Alan Cripe, Dr. Young and others conveyed not only their views about the environmental damage in store if the parkway were built but also the heavy financial risk. One big flood during construction, it was argued, would send millions of tons of fill dirt to Newport News. Also, tolls, in their calculation, would never pay for the cost of this section of roadway. This two-pronged argument, devastation of the environment and financial loss, continued through the events in the days that followed.

A part of Discover Day was the solicitation of signatures on a petition to the Richmond City Council; the event produced an estimated two hundred signatures. The concluding lines on the petition no doubt made and impression: "The beauty of this area is obvious to all: ONCE DESTROYED IT CAN NEVER BE REPLACED!"

MEETING WITH CITY COUNCIL

A headline in the *Richmond Times-Dispatch* on November 18, 1970, read "Parkway Plan Held No Threat to River." The headline drew "protests of outrage" because the story reported on a joint meeting between the Richmond City Council and the Richmond Scenic James Council in which the vast majority of presentations enumerated the perceived threat to the river. This meeting, held in city council chambers in "Old City Hall," was not a public hearing in the formal sense, but an "open meeting of the Richmond Scenic James Council with City Council present."[57]

Mayor Thomas J. Bliley Jr. opened the proceedings recognizing that this "meeting of council has been called at the request of the Richmond Scenic James Council who wish to come before council to present their concern of possible harm to the River if the proposed James River Expressway link of the Richmond Metropolitan Authority is constructed in its present location."[58] Richard Obenshain, attorney for the Richmond Scenic James Council, introduced the purpose of the meeting, noting that the "thing that has brought this group together as many of you probably already know is a passionate attachment and love for the great river which runs through our city." He then turned the meeting over to Dr. Young, president of the Richmond Scenic James Council, who served as moderator. "The ultimate goal of the Richmond Scenic James Council," Young opened, "is to preserve the natural beauty and ecology of the James River, its shoreline and its many wilderness islands as it courses through the fall line in the metropolitan Richmond area. Precious few cities in America can boast a river of such historic and scenic charm."

Dr. Jack P. Andrews, one of the principal speakers, described features of the expressway plan, with John Hurst pointing to a map as he talked. Dr. Andrews's vivid description of the expressway plan emphasized the disruptions to the natural features of the river that the Richmond Scenic James Council members feared. After this basic description of the RMA plan and his perception of its destructive impact, Dr. Andrews ended his presentation with a rhetorical flourish: "This great scenic attraction in the heart of our capital city has withstood the ravages of time, flood, drought, snow, and ice; but it cannot withstand the bulldozers."

The lineup of speakers continued. Alan Cripe, transportation consultant, argued against the proposed expressway in pragmatic terms, using detailed calculations to claim that there was no traffic need for the expressway and that it would cost the taxpayers of Richmond dearly, far more than tolls could make up. Several biologists from the University of Richmond, Virginia Commonwealth University and the National Audubon Society made presentations at the forum, detailing the damages they feared the expressway would impose. One labeled the prospect an "environmental and ecological calamity of the worst kind, an ill-advised move for which Richmonders and Virginians would pay dearly for generations to come."

Garden clubs were an important component of the Richmond Scenic James Council, and Mary Frances Flowers, member of Boxwood Garden Club and president of the Garden Club of Virginia, contributed to the city hall symposium. Leaders from several other organizations reported the opposition of their boards to the proposed expressway, including Rich Koster, the Mid-Atlantic Council on Pollution in the Environment; Robert Hicks, Southampton Citizens Association; Elsie Elmore, Board of the League of Women Voters; and Samuel A. Anderson III, Richmond Metropolitan Section of the Virginia chapter of the American Institute of Architects. A. Howe Todd, director of planning, expressed the opposition of city planning staff to the proposed plan with the recommendation that a far more limited roadway be considered.

After brief discussion of copies of various documents, Mayor Bliley introduced Charles A. Taylor, chairman of the Richmond Metropolitan Authority, who was the primary spokesperson for the RMA position. Taylor was quite conciliatory in tone, saying that "everything possible and reasonable should be done to protect the scenic beauty of the James River Basin and the ecology of the area" and that he was willing to cooperate with "any group or organization which is interested in these values." Defending the RMA plan, however, he noted that the parkway had been an approved part of plans going all the way back to the City Master Plan of 1946 and RMA staff had already made more than 150 appearances before various civic groups.

Taylor's was the last of the formal presentations. The meeting was then opened for questions and answers and discussion. The essential features of the discussion were efforts by some city council members and the leadership of the Richmond Scenic James Council (1) to get assurance that the sale of bonds would not commit the Richmond Metropolitan Authority to constructing the Riverside portion of the expressway and (2) to convert the RMA's promise to work with concerned groups into a commitment not to construct the Riverside Parkway. Councilman James Carpenter pressed vigorously for assurance on both of these points.

The effort to pin down the RMA went through a number of additional iterations, with flexibility and cooperation offered but deletion not promised. Carpenter recommended that a smaller meeting be held the following week with participation by the Richmond Metropolitan Authority, city officials and the leadership of the Richmond Scenic James Council. Mayor Bliley agreed to set up such a meeting.

MOP-UP

The meeting promised by Mayor Bliley at the conclusion of the hearing was convened in the mayor's chambers on November 23, less than a week later.[59] The RMA agreed to remove reference to the Riverside Parkway in the bond prospectus and to provide written evidence of this action. Nevertheless, attorney Richard Obenshain reinforced the insistence that the unwanted references to the Riverside Parkway in the proposed official statement be deleted. He delivered a letter to Mayor Bliley letting him know that the draft still contained "numerous positive statements which would lead a prospective bond purchaser to conclude the RMA System, when completed, will contain a Riverside Parkway." He asked the mayor and his colleagues to turn their "persuasive powers upon the Richmond Metropolitan Authority to obtain the immediate revision of the offering document."[60]

Despite the sense of victory in December, concerns continued to be felt. Plans for the expressway and the draft of the City Master Plan continued to show an interchange at the south end of Powhite Bridge that the Richmond

Scenic James Council feared might be a stepping stone to later revival of some version of the Riverside Parkway. Ultimately, this feature was removed from the City Master Plan, the parkway was removed from the RMA's bond solicitation and this stretch of road was never built.

PERSPECTIVE

How should one interpret the discontinuation of the plan for Riverside Parkway? One of the earliest activists against this leg of the expressway asserted strongly that he is "convinced that the economics stopped the riverside leg and not concern for the river."[61] The more "romantic" argument is that concern for the river expressed through organized citizen protest not only won the day but also led to an expanded and continuing appreciation for the river and concern for the environment.

One can reasonably conjecture that citizen protest alone, no matter how passionate, would not have stopped the Riverside Parkway. After all, there was considerable protest surrounding other segments of the expressway, including an effort to prevent the destruction of the remains of the Kanawha Canal in downtown Richmond. Well-organized protest, including a legal challenge, did not stop that stretch of the expressway. Perhaps, however, the citizen protest against the riverside segment added weight to an existing suspicion, if not recognition, of its economic weakness. Perhaps the proverbial straw broke a relatively weak camel's back. Clearly, the protesters proclaimed at every opportunity that the project was economically risky.

One can also ask, had there been no protest at all, would the inherent economic weaknesses and risks have stopped the Riverside Parkway? It is possible that further engineering and traffic studies would have led the RMA to discontinue this leg. The tenacity with which the RMA held to its original route and the reluctance with which it abandoned the possibility of future construction as conditions changed would seem to indicate it would not have arrived at this decision independently.

In conjecturing about this piece of river history, one can also ask: Did the fight against the Riverside Parkway build a level of concern and support for the river that would not have occurred without it? The answer seems to be in the affirmative. The James River in Richmond, as the Ringle article quoted earlier points out, had largely been neglected after it lost its industrial function. Still, there is little doubt that multiple forces were at work to increase environmental awareness and appreciation for the river. The direct line of activism from the Richmond Scenic James Council to the more than forty-five years of the Falls of the James Scenic River Advisory Committee and its accomplishments suggests that the expressway fight did play a part.

The Richmond Scenic James Council continued as an organization for a number of years, but its legacy, in addition to whatever effect it had on stopping Riverside Parkway, is its spawning of the scenic river committee that has endured. Several of the members in the parkway fight, along with attorney George Freeman, worked with the initially reluctant city and the Virginia General Assembly to establish the Historic Falls of the James Scenic River Advisory Committee in 1972, a special scenic designation. The "special" label was removed and the committee brought under the Virginia Scenic Rivers Act in 1984. That organization, with some variation in name and structure, still functions today. Several members of the expressway fight, including John W. Pearsall and Louise Burke, served for many years. Dr. R.B. Young chaired the committee for thirty years, from its inception in 1972 until 2002. Richard D. Obenshain was killed in a plane crash during his run for the U.S. Senate in 1978.

The activism stimulated by the proposed Riverside Parkway was an early indicator of broader changing attitudes toward the James River in Richmond. The proposed location of the parkway had been selected and defended on the grounds that it was a floodplain and thus would not require removal of many structures, was of little value for other uses and would cost less than other properties to acquire. Those arguments have begun to lose their power. The change in values and attitudes was part of a national awakening of environmental awareness, but it also had a distinct local flavor. The James River in Richmond was beginning to be valued for environmental, aesthetic and recreational reasons rather that the utilitarian purposes of the past. It signaled a shift in what is to be valued. The opposition to and defeat of the proposed parkway was both an indication of the increasing interest in the James River and a source of environmental activism that continued well after the expressway controversy was a vague and distant memory.

———

The shoreline saved from the proposed expressway became a popular part of the James River Park System, which was already in the development stage while the expressway battle was unfolding.

4

PUBLIC ACCESS

BUILDING THE JAMES RIVER PARK SYSTEM

It is difficult to explain to people now that before the 1970s,
we couldn't get to the river without trespassing on somebody's property.
—*a river user, 1995*

A Little Bit of Wilderness in the Heart of the City.
—*motto of Friends of James River Park*

At one time, the entire shoreline of the James River in Richmond was private property, and access required ownership, permission or trespass. Testimonials abound regarding the lack of access to the river prior to the establishment of the park. "They can see the river but they can't touch it" said a mother in 1970 about her children's inability to get to the river.[62] In 1995, looking back, a river user said, "One of the best developments in my lifetime has been the removal of barriers along the James River. It is difficult to explain to people now that before the 1970s, we couldn't get to the river without trespassing on somebody's property."[63] Getting to the river meant crossing private property, rail lines or canals or some combination. The James River Park has opened access and has provided protection from riverside development that floods and rail lines do not offer.

The dedication of the first section of the James River Park occurred on October 11, 1970, the same day as the public display of opposition to the Riverside Expressway farther upstream. Both events were important in making the river what it is today. FitzGerald Bemiss, chair of the Virginia

Commission on Outdoor Recreation, spoke, calling the park "the first step at getting to our river—a river that has been abused so long."[64]

The current park "system" has been assembled through the acquisition of parcels one by one, a process that is still ongoing. The success of the park can be judged not only by its growth in size but also by its popularity. A 2015 count showed that the James River Park is the "most-visited place in Richmond—by a wide margin," and visitor numbers have continued to grow.[65] The park has enhanced the image of the river, and the river has enhanced the image of the city.

EARLY NOTIONS AND PARK FORMATION

First steps toward the development of the park occurred many years earlier than the 1970 dedication. It appears that this early park interest was not to use the river for fishing, hiking, swimming or boating but to view the river from afar. The 1946 City Master Plan captures this interest and laid the foundation for action to follow: "Exceptional opportunities exist for the development of an outstanding riverfront drive alongside the south side of the James River....Property lying between the drive and the river should be brought under public control in order to protect this property from uses which might be detrimental to the use of the drive for pleasure driving." City planner Garland Wood is also on record supporting the city's interest in providing a view of the river: "Indeed, on all of the high level land between Riverside Drive and the Southern Railway parking areas should be provided so that the motorist can enjoy the view without obstructing traffic....A few benches and picnic tables placed around these areas in appropriate places would afford an excellent vantage point from which the scenery could be enjoyed."[66]

Although visual access to the river was the theme in the 1940s, physical access soon became part of the agenda. Indeed, both physical and visual access to the river have been continuing issues well into the twenty-first century. A major step in the assembly of the James River Park System occurred in 1966 when the city accepted deeds to a number of islands on the south side of the river conveyed by Charles J. Schaefer and John W. Keith. Schaefer acquired title to the nearly two-mile stretch of islands by paying delinquent taxes.[67] A plaque prepared by the Friends of James River Park now acknowledges their contribution.

During the late 1960s, planning for the park was well underway, and pieces began to fall into place for its establishment. In 1967, the Virginia Commission on Outdoor Recreation indicated that it would recommend the financing of Richmond's park plans for the James River, and in 1968 federal funding became a reality with the awarding of a federal Land and Water Conservation grant to be matched by state and local funds.[68] An important part of the story for the next several years was funding achieved by cobbling together state and federal grants usually matched by city funds. Primary funders were the U.S. Bureau of Outdoor Recreation and the Virginia Commission on Outdoor Recreation. Funds were used both to acquire additional properties and to develop facilities and make improvements in the park.

A master plan for the park prepared by landscape architect Stanley W. Abbott was transmitted to the city in 1968. Abbott's plan was comprehensive and ambitious, covering Williams Island, the Pump House parts of the Kanawha Canal, North Bank and Belle Isle. When the park

One of the bridge-staircase walkways that brings visitors over railroad tracks to the park. *Photo by author.*

was dedicated in 1970, however, it included only the Main Section—two miles along the south bank of the river between "Nickel Bridge" and Lee Bridge, two parking lots with pedestrian bridges over the rail tracks and walking and biking trails.

WHAT KIND OF PARK?

An issue that arose early and continues to this day is the "character" of the park. The character question was (and is) about how "natural" versus how "developed" the park should be. Although Abbott endorsed the concept of maintaining a natural environment, he proposed amenities far greater than later "natural" visions of the park would allow. Williams Island, for example, would have boating concessions, picnic areas, a shelter pavilion, a comfort station and more. Belle Isle would be connected by a monorail and hold two restaurants, a central fountain plaza, boat concessions, trails, a swimming pool, a picnic area and museums. The plan would "Consider Belle Isle the show piece of the James; downtown pleasuring ground for Central Business District and Capitol."[69]

Concerns were raised almost immediately after release of the park plan. Abbott responded vigorously in a letter to Jesse Reynolds, director of the department of recreation and parks, to "the charge that we are overdeveloping the James."[70] Abbott pointed to the "#1 guideline" from his plan: "The glory of Richmond's James is the comparative naturalness of the setting. Consider preservation of the existing green sheath of wooded banks and wooded islands as a goal of paramount importance. In line with this reasoning the master plan limits major recreational development to areas where natural conditions have been altered previously."[71]

His letter underlined for emphasis "where natural conditions have been altered previously." The development on Belle Isle, for example, would be east of the Lee Bridge, which is "an area of industrial condition today—not of natural condition." Likewise, the development on Williams Island would be in "an area to be greatly changed by construction of the proposed metropolitan expressway." Alternatively, areas that were still natural would remain natural.[72]

Despite Abbott's contention that only areas already disturbed would be developed, the criticisms and arguments to "leave it natural" continued.

One early note for a natural approach was a letter to the editor by Charlotte Opsahl. She expressed concern about even creating a park because "once the planners are given a lovely natural area to use as a park, they feel they must immediately add wide paths, bridges, picnic pavilions, benches—in short, all the trappings of civilization." She continued her argument: "This is a plea for the planners to leave at least a portion of it for those of us who don't mind getting our feet wet, and ducking through the undergrowth to reach spots where we can enjoy some measure of seclusion and experience the remaining scraps of wild beauty along the riverside."[73]

The "keep the park natural" theme was emphasized in a report prepared for the city in 1973 by Joseph J. Shoman of American Conservation Planning Associates. "The primary purpose of the James River Park," the plan states, "should be the preservation of the natural values inherent to the river; the secondary purpose should be to provide limited outdoor interpretation and education and low-keyed, non-consumptive, forms of outdoor recreation along the river banks, meadows, woods, and in the river itself."[74] The report argues that any overdevelopment from 14th Street to Bosher's Dam would be "a grave mistake—an ill-advised action

One of the many "natural" spots in the James River Park System. *Photo by author.*

Another natural scene, even more striking with early fall foliage. *Courtesy of Scott Weaver.*

that would haunt the decision makers for many generations to come."[75] In addition to Abbott, Opsahl and Shoman, citizen leaders such as Louise Burke, R.B. Young, Robert Hicks and John Pearsall added their influence to the naturalist perspective.

The focus on keeping the park wild and natural was given a boost by nature itself in the form of the Tropical Storm Agnes flood during the summer of 1972. Among other influences, the flood convinced planners to remove plans for bridges to and between islands. Nature itself was reinforcing the decision to keep the park "natural."

It seems that the "character" of the park question was answered early and has for the most part been maintained—although, as longtime park manager Ralph White would likely attest, it has taken constant vigilance. White displayed unflagging commitment to maintaining a "natural" park unblemished by vendors or amenities that do not seem to fit a natural environment. At the same time, he did not try to protect the park by keeping people out. Rather, his belief was that visitors become enthusiasts who in turn become supporters, volunteers and protectors of the park.

A variety of park management strategies have been used to allow visitors but still keep the park as natural as a frequently visited landscape can be. Access is made easy in some areas, difficult in some and off-limits in still others. Park trails are restricted to walking or pedaling, with no motorized vehicles allowed. Parking lots are kept at moderate size to reduce crowding, and except for Ancarrow's Landing in the tidal reach of the river, only hand-launched craft are permitted (enforced by the design of the launch sites, not enforcement action). No commercial services are allowed in the park—no restaurants, food vendors or vending machines. The exception is outdoor programming. Park rules allow nonprofit and commercial companies to use the park for pleasure and training activities, a practice that has caused some conflict from time to time (see chapter 6).

ASSEMBLING PIECES

The years following the dedication of the Main Section of James River Park saw the announcement of additional funding and the acquisition of new properties on both the north and south sides of the river. Primary developments were on the south side, but some attention was also given to the north side of the river, with a grant from the Virginia Commission on Outdoor Recreation designated for Phase II development on the north bank.[76]

BELLE ISLE

After the hydropower plant was closed in 1967 followed by Old Dominion Iron in 1972, the city purchased Belle Isle, thus making plans for its use as a park possible. For a period of time after industrial use of the island ended, access to Belle Isle was limited. A variety of ideas were suggested for reaching the island, including a mini-rail from the central business district, a bridge and parking for motorized access and a passenger ferry. Until the new Lee Bridge was constructed in 1985 and the pedestrian bridge slung underneath, however, access to Belle Isle was limited to paddlers or those who walked through the rock garden river channel on the south side of the island, which is usually kept relatively dry by the dam from the tip of the island to the south shore. Paddlers through Hollywood Rapid in those

A view of Belle Isle in 1965. Note the industry on the island and the main flow on the south rather than the north side of the island. *Courtesy of Cabell Library Archives, Virginia Commonwealth University.*

Transforming Belle Isle for recreational use required cleaning up the clutter of abandoned industry. *Courtesy of Venture Richmond.*

days were said to "risk" being mooned or sighting nude sunbathers on the riverside rocks. Belle Isle has come to be one of the most highly visited segments of the park system.

PONY PASTURE (NOW PONY PASTURE RAPIDS)

The Pony Pasture area (it did pasture ponies at one time) on the south side of the river had been used as a de facto park for years, not always to the satisfaction of nearby residents. Its transformation as a designated part of the James River Park began in 1972. The expressway proposal that would have decimated that part of the riverbank had been turned away, and the city annexed the area from Chesterfield County in 1970, but one roadblock still stood in the way of making Pony Pasture a park. This riverfront property was privately owned, and the owner of the most desirable plot, Stern and Arenstein Enterprises, was proposing to build an upscale apartment complex on the site.

Richmond City Council voted unanimously to reject the apartment proposal and to initiate a purchase process.[77] The apartment proposal, which also would have donated twenty-four acres for a park, had numerous obstacles in addition to the city's interest in ownership. These constraints included traffic congestion because of the narrow streets in the area, floodplain issues, difficulty of fire and emergency access and violation of the subdivision ordinance.[78] Additionally, it is likely that strong community opposition would have arisen if the plan had moved forward. Some residents, with a touch of cynicism perhaps, suggested that the apartment proposal was not serious but rather a ploy to raise the purchase price of the property.

The city shortly thereafter appropriated funds and acquired the Stern and Arenstein property (fifty-one acres), the Schultz property (four acres) and the Blankenship property (forty acres), all with river frontage. Together, along with a small section already owned by the city, these properties stretched along the river from the bend in Riverside Drive downstream to the Willow Oaks Country Club. The western portion, beginning alongside the rapids, became Pony Pasture Park (now called Pony Pasture Rapids), and the eastern section was named The Wetlands, both part of the James River Park System.

All was not peaceful in Pony Pasture, both before and after it opened as a park in 1981. As the story goes, this was a favorite hangout for Hell's Angels, Pagans and drunk teenagers. Severe measures were considered to

Pony Pasture has been a popular summer spot for decades. *Photo by author.*

address the problem as a presentation by J. Robert Hicks to Richmond City Council indicates. "Closing the Pony Pasture parking lot and erecting a chain link fence," Hicks argued, are "extreme reactions which are expensive and deprive the citizens who wish to use the park properly of one of the most valuable and beautiful inner-city natural areas in the eastern U.S."[79] It was not these measures or a crackdown of law enforcement that turned the tide. Rather, encouraging families to use the park was the remedy. "Pregnant ladies and screaming kids ruined the location for the tough guys" said Ralph White in a recent conversation.[80]

HUGUENOT WOODS (NOW HUGUENOT FLATWATER)

Simultaneously with the purchase of Pony Pasture and The Wetlands, a process was underway to acquire acreage on the south side of the Huguenot Bridge approximately two miles upstream of Pony Pasture.

The property owners, May (seven acres) and Carneal (twenty-one acres), were reluctant to sell to the city for purposes of a park because they were concerned about undesirable uses and activities if the property were opened to the public. John Pearsall, Dr. R.B. Young and others acted as brokers for a two-step acquisition by the city. A key instrument in this brokering process was the Historic James Greenbelt Corporation initiated by John Pearsall and chartered in December 1972 as a nonprofit corporation for the purpose of facilitating public access to the river. After building in deed restrictions regarding the types of uses and activities that could take place, thereby placating the concerns of the owners, the Greenbelt Corporation purchased the property. Additional use restrictions were added, and the corporation then sold the property to the city for the price it had paid. Huguenot Flatwater now is a busy launch site for the calm water both up- and downstream, and fundraising is underway to build a universal access ramp for handicapped paddlers.

ONE EXPANSION NOT SO SUCCESSFUL

As the fight against the proposed Riverside Parkway was coming to a close in 1970, the Richmond Scenic James Council announced that it had begun plans for the development of a rustic trail for hiking and cycling along Riverside Drive between Huguenot Flatwater and Pony Pasture Rapids. The vision of these park proponents was to make both Huguenot and Pony Pasture part of the park system and connect them with a walking path.[81] This would, in effect, expand the park by establishing continuity between Huguenot Woods and Pony Pasture.

On more than one occasion since that initial effort in the early 1970s, attempts have been made to establish a trail or walkway along this two-mile shoreline alongside Riverside Drive. On each occasion, controversy erupted around the issue and the project turned aside. The residents who opposed the path made the claim that privacy and property rights trumped public access to the area. The proponents, including several residents from the same area, argued that the public should have the opportunity to enjoy this reach of the river without risking injury on the narrow roadway. The resistance to the path may be explained, at least in part, by the historical experience in the neighborhood.

The central draw in the area is Pony Pasture Rapids, a section of Class II whitewater that is dotted by large boulders easily accessible from shore in moderate to low water levels. Especially on warm summer weekends, teenagers and young adults (among many other users) flock to the area. This section of the shoreline is now part of the James River Park System, but the summer rock-hopping and swimming ritual began many years before the area was established as a park. These users of the river were interested in more than just wading and swimming in the "town swimming hole." It also was a place for parties, sometimes loud. And often there were more cars than parking places, so cars were parked along a narrow roadway and even in the driveways of residents. The "parking and partying" problem was making it more difficult to bring about the initial effort to establish the walking path and open access to the river along Riverside Drive for fishing, hiking and cycling.

In the early 1990s, a new initiative to establish a walking trail along this section of the river was launched. The city contracted with Carlton Abbott, son of the landscape architect who had designed the original James River Park facilities, to design a trail along this two-mile stretch. He developed a report with a design for a ten-foot-wide bikeway/walkway.[82] In 1992, a meeting of Riverside Drive riparian owners at the home of John and Laila Pearsall was convened to discuss the proposed Riverside Trail designed by Abbott. The plan and the meeting provoked a storm of opposition; John

Riverside Meadow Greenspace, a two-acre walk-in park, was donated to the city with use restrictions in consideration of nearby neighbors. *Photo by author.*

Pearsall quipped that he "thought they were going to be lynched that day." He attempted to reduce opposition to the plan by suggesting that restrictions similar to those imposed on properties donated through the Historic James Greenbelt Corporation be attached to the Riverside Trail. Those opposed were not persuaded. Attempts several years later to initiate less ambitious plans were similarly turned aside. Some residents along Riverside Drive also objected to the Pearsall donation of 1,500 feet of river frontage to the city, but multiple use restrictions facilitated acceptance. That property is now a "walk-in" park called Riverside Meadow Greenspace and is part of the James River Park System. Other properties have been added to the park over the years, and if the new park master plan is followed, the park will continue to grow in size.

STAFFING

According to supporters, the James River Park has been underfunded and understaffed throughout most of its history. At least as early as 1975, the Falls of the James Scenic River Advisory Committee was making the case for expanding the park's budget and for more staff.[83] The perception that the park was not adequately funded by the city continued year after year. At times, even structural options were suggested such as creating a regional parks foundation (on the Maymont model some suggested) with the James River Park as its centerpiece.

With multiple demands on its resources, support vacillated for the James River Park System by its home agency, the Richmond Department of Recreation, Parks and Community Facilities. The need to devote attention to inner-city youth and inner-city problems consumed much of the department's resources in the form of ball parks and swimming pools. Nor were the attitudes of the city bureaucracy always supportive of the river park that was emerging. As one department head is said to have asserted, "who wants to have a bunch of weeds and snakes?" Moving water was thought by some to be a hazard in public recreation and an unnecessary risk. Even staff were thought to be at risk of injury working on river rocks and forested or brushy shoreline.

One early call by advocacy groups was for a park naturalist. That request was answered in 1980 with the employment of Ralph White.[84] Over time, White's role morphed from naturalist to park manager and naturalist (and maintenance worker and ranger and public relations officer). At times, his staff included one or two full-time employees and seasonal workers, and at other times he alone was park staff. The gap in personnel funding was filled, at least in part and at least some of the time, by volunteers, donors and White's willingness to work well beyond his job description. The magnetic draw of the James River as well as White's power of persuasion attracted thousands of volunteers over the years. Boy Scouts have completed dozens of projects; schoolchildren have spent time picking up litter in return for nature programs (one hour of pickup buys one hour of nature program); businesses and environmental groups have contributed funds for signs, brochures and materials. Volunteers continue to come as individuals, as groups from schools, universities and businesses and in the form of support organizations such as Friends of James River Park, James River Outdoor Coalition (JROC) and Mid-Atlantic Off-Road Enthusiasts (MORE).

White's passion for the park and his vision for it were at times a source of conflict with the "bureaucracy," and White occasionally was "called on the carpet" by his superiors in city hall. Perhaps it is this passion that has made him an iconic figure among river and park supporters (and recipient of multiple local, regional and national awards for his work).

White retired in December 2013; Nathan Burrell, former intern, seasonal worker and park and then city-wide trails manager, became the new park manager, with the position now upgraded to park superintendent. Despite worries by some park supporters that the park would fall on hard times without the iconic White at its head, Burrell has been up to the challenge. His style is different, but he proved to be a strong leader who effectively continued the vision for the park while simultaneously expanding city administration support and park activities. Both park staff and programming increased under Burrell, who has moved up to a new position. Bryce Wilk, also an experienced park professional, was hired as James River Park superintendent in 2018, in time to participate in the park master planning process.

WATER, GRANITE AND DIRT

For outdoor activists, the James River Park offers three resources: water, granite and dirt. The water for paddling, swimming and fishing is obvious. But the dirt (in the form of trails for hiking, running and cycling) and the granite (in the form of climbing sites) also have their enthusiastic users and supporters.

WATER: PADDLING FACILITIES
(AND THE SEARCH FOR A DOWNTOWN TAKEOUT)

The establishment of James River Park gave a major boost to river access for paddlers. The park has provided boat hand-launch sites at Huguenot Flatwater, Pony Pasture Rapids and Reedy Creek as well as a motor launch at Ancarrow's Landing. Each of these offers a different boating experience: Huguenot is a flatwater site where most paddlers put in and take out at the same ramp; Pony Pasture is the preferred entrance to an intermediate whitewater trip with takeout at Reedy Creek. Reedy Creek is the launch site for a trip through the more challenging downtown rapids ending typically at 14th Street Takeout, and Ancarrow's Landing provides motorboat access to the tidewater section of the river. Thanks to nature's design and well-placed access points, the park offers something for every skill level and type of boating.

A gap in this comprehensive paddling layout arose at the end of the 1990s when a takeout was no longer available at the natural conclusion of most downtown whitewater paddling trips. In the 1980s, a takeout with necessary parking was located at Reynold's Metals Company. The floodwall disrupted that location, leaving no place to park shuttle vehicles. Mayo's Island then became the takeout of choice. It served well, including the central takeout for the 1998 Whitewater Open Canoe National Championships and even featured a retail paddling shop for a short time. However, ownership and rental patterns changed, removing Mayo's as a takeout location.

Although many in Richmond bragged about having the nation's finest urban whitewater, a frequent quip at the turn of the twenty-first century was that "you can put on, but you can't take off."[85] Paddlers considered this the river version of the Kingston Trio song "man who never returned…forever beneath the streets of Boston."

Several organizations participated in the search for a suitable takeout, both by pressure on the city and by walking the riverbanks to locate a spot that might work. The James River Outdoor Coalition, the Richmond Whitewater Club and the Falls of the James Scenic River Advisory Committee all took up the cause. Letters were sent to the mayor, city manager and council members. Meetings were held and phone calls made, all seemingly to no avail.

The location identified in the riverbank search with the most potential was a vacant lot at the northeast corner of the 14th Street (Mayo's) Bridge. Much of that property was owned by Norfolk Southern, but word was that the city was planning to acquire it. If the city did purchase it, the site would be near perfect for a takeout—that is, if the city did not use it exclusively for public parking. The river groups developed a proposal that would make the site multiuse to include public parking and dedicated parking for paddlers, a miniature park along the river and a paddling takeout. The proposal languished.

With hopes for this "grand design" diminishing, advocates continued to explore the riverbanks. In one of those explorations, the property adjoining the Norfolk Southern lot, part of the department of public utilities wastewater operation, seemed to offer some potential. It was small, somewhat cluttered and adjacent to the Shockoe CSO Gate. But perhaps, it was thought, it could be made to work.

Conversations followed with Robert Steidel, DPU deputy director at the time, and Ralph White, park manager. Shortly after, at what seemed like warp speed considering earlier frustrations, a plan was underway and a memorandum of understanding between departments was signed. Not long after, construction of a ramp and stairs was underway. With

Staff and volunteers spend hundreds of hours constructing the elusive 14th Street Takeout. *Courtesy of rich young.*

14th Street Takeout after completion. *Photo by author.*

the leadership and long hours of labor from Peter Bruce and Nathan Burrell on the park staff, along with hundreds of hours of volunteer help, especially from the James River Outdoor Coalition (JROC), and with financial support from Friends of James River Park and JROC, the 14th Street Takeout was completed in 2005. JROC continues to provide much of the maintenance. As the 2012 Riverfront Plan attests, the site deserves expansion and may yet receive it.

DIRT: A GROWING TRAIL SYSTEM

Trails have been an important part of the James River Park from the beginning, and the network of trails, both inside and outside the park, has grown considerably over the years, especially since 2003 when Nathan Burrell was hired as park trails manager and then made city-wide trails manager. When Burrell was hired as park superintendent, the city-wide trails manager position was continued and filled by Michael Burton. So, trails are now a city-wide project. James River Park trails increasingly connect with trails that are not part of the park system: Virginia Capital Trail, Slave Trail, Forest Hill and others.

The growth of trails has been the result of a productive relationship between the park and several active organizations. There are some "use paths" generated by natural traffic, but most trails in the park are carved out purposely for hiking and biking. Miles of these trails have been constructed by volunteers in organizations including the Richmond affiliate of Mid-Atlantic Off-Road Enthusiasts (rvaMORE), Richmond Triathlon Club and the James River Outdoor Coalition, International Mountain Bike Association, Richmond Sports Backers and James River Hikers. Volunteers and park staff have spent thousands of hours building and maintaining trails. Multiple mountain biking and running events like Dominion Riverrock and XTERRA competitions have been held on the complex of trails.

Just as paddling opportunities in the park are suitable for different skill levels and interests, so are the trails—from level and easy to steep and demanding, for dog-walkers and mountain bikers, for runners and bird watchers. Although there are opportunities to venture out and return on the

Miles of trails parallel and cross the river for hikers, runners and mountain bikers. Several connect with trails outside the park. *Courtesy of Scott Weaver.*

same route, there are multiple loops that do not require retracing one's steps. Some loops are short—less than a mile—and some as long as thirteen miles. Some cross the river, and many do not. Trails have a variety of official and unofficial names: Buttermilk, Buttermilk Heights, North Bank and "poop loop" (near the wastewater treatment plant). In 2007, two trails named after early river advocates were dedicated in Pony Pasture and The Wetlands: Louise Burke Nature Trail and R.B. Young Riverside Trail.

A relatively new addition to the trails network is the Mountain Bike Skills Area opened in 2012 on Belle Isle under the Lee Bridge. Here new riders and children have the opportunity to build skills in a controlled environment before heading to the trails.

GRANITE: CLIMBING

Topography does not allow the diversity of opportunities for climbing that are available for paddling and hiking/running/biking. Nevertheless, three locations do provide a place for climbers to hone their skills: the Manchester Climbing Wall and two locations on Belle Isle. The climbing wall is an abandoned abutment for the Richmond and Petersburg Railroad Bridge equipped with climbing anchors; the quarry sites on Belle offer a number of climbing routes on the walls of old granite quarries.

The Manchester Climbing Wall inherited by climbers from the long-abandoned Richmond-Petersburg Railroad Bridge. *Photo by author.*

The granite climbing wall at the quarry site on Belle Isle. *Photo by author.*

CONSERVATION EASEMENT

A worry of many park advocates at one time was that the draw of real estate development and tax dollars would lead to the sale or inappropriate use of portions of the James River Park. This fear was alleviated when city and state officials signed an agreement placing much of the James River Park under a conservation easement designed to permanently protect the park from sale or development. Governor Timothy M. Kaine and Mayor Dwight C. Jones signed the final document in a ceremony on Brown's Island in May 2009.[86] The route to that signing ceremony had a few twists and turns.

After convincing the city to establish Bandy Field as a city park in 1998, Dr. Charles Price set out to establish a conservation easement on four properties: Bandy Field, Crooked Branch Ravine, the Lewis G. Larus Tract and the James River Park System. With the leadership of Dr. Price and the involvement of multiple organizations and individuals, city council was coaxed into adopting a resolution in May 2000 directing the office of the city attorney to prepare an ordinance to protect the four sites with a conservation easement. The city administration, however, argued against the use of conservation easements with the reasoning that adequate protection existed and that easements would unduly tie the hands of future officials, so the city council backed away from pursuing easements at that time and the issue was dropped. (Tying the hands of future officials, of course, is a purpose of a conservation easement.)

The matter seemed settled, with no easement likely, until the specter arose of the city selling Great Ship Lock Park and part of Chapel Island to a developer as an incentive for building on adjoining property. Pushed by a coalition of groups, the idea of a conservation easement was resurrected and placed back on the agenda. After a good deal of discussion, the decision this time around was to focus on the James River Park, in part to establish a model that later might be applied to other parks.

Council members Kathy Graziano and Bill Pantele actively participated with the coalition of citizens working for the easement. Graziano, in particular, became a champion for the cause. On October 18, 2005, the city's land use committee supported a resolution directing the city attorney to prepare conservation easement documents for the James River Park. The issue was then on the way to city council, and supporters were urged to write, call, visit and email council persons. An editorial in the *Times-Dispatch* supported the proposed easement: "Mrs. Graziano's proposal affords the city a chance to do itself a favor. Conserving the park through an easement would safeguard a Virginia treasure for future generations, and protect an irreplaceable aspect of the city that makes Richmond a unique and wonderful place to live."[87]

At the October 24, 2005 council meeting, Resolution 2005-R210 passed unanimously with all council members present signing on as co-patrons. The resolution alone did not establish an easement; it simply authorized the detailed work required to produce a conservation easement document. Both city staff and citizens, especially John J. Zeugner, spent many hours developing that document. Two key city agencies, in addition to the city attorney, were the department of parks, recreation and community facilities, of course, and the department of public utilities. With responsibility for water supply and waste disposal, DPU is integrally involved with the river and the title holder for a number of parcels along its banks. The detail work was carried out through the Enrichmond Foundation (formerly the Richmond Recreation and Parks Foundation) with Zeugner as project director. Three and a half years later, the document was ceremoniously signed and filed in court.[88]

CONTINUING TO GROW

The conservation easement marked an important point in the ongoing development of the James River Park System, but by no means was it the final step. The park continues to grow both in the territory and facilities it includes, the programming it offers and the number of visitors it attracts. Led by the fundraising efforts of Friends of James River Park, a park master planning process, reminiscent of its formative years, has been completed, and the resulting plan was adopted by city council on January 27, 2020. Much is on the agenda for the park. The master plan suggests many new opportunities, including physical expansion, greater connection with trails outside the park, revenue enhancement and much more. It also addresses old problems like invasive plant species, parking shortages and potential overuse.

PERSPECTIVE

The James River Park System plays an essential role in making the river a valued centerpiece of Richmond and an amenity accessible to the public from Richmond and beyond. Without the park, the river would still be, in the words of the frustrated mother, a place her children can see but can't touch. Even seeing the river would be more restricted without the park. The park has contributed immensely to the favored status the river now holds in Richmond. The park itself has been a recipient of significant awards as well as a major contributor to other accolades the city has received. The James River Park is now the most visited place in the city, approaching two million visits per year, a level of popularity that raises the question of overuse.

The park has grown over the years from the initial "main section" on the south side of the river to include Huguenot Flatwater, Pony Pasture Rapids, The Wetlands, Belle Isle, Ancarrow's Landing, Great Ship Lock Park, Byrd Park Pump House, T. Tyler Potterfield Memorial Bridge, among others. Other units, Williams Island, 14th Street Takeout, much of the Slave Trail and Floodwall Park, are managed as part of the park while technically on the inventory of other city agencies. The cooperation with the department of public utilities has been especially positive and important.

The park that has come to be is quite different than the early plans that would have placed a boat concession and picnic tables on Williams Island and restaurants and a swimming pool on Belle Isle. The argument for a "natural" park by early advocates and park management over the years (and no doubt the high cost of many of the amenities proposed early on) have made it a much different place than some early planners envisioned.

From advocates' point of view, historically, the park has been underfunded and understaffed. Even keeping the park natural requires funding and staffing. During much of the history of the park, city hall had other priorities and in some cases near disdain for the James River. The deficiency of underfunding and understaffing has had a positive side—the need for volunteers. Park leaders have encouraged the participation of volunteers, both individuals and groups, which in turn has helped create a sense of "ownership" and a cadre of park enthusiasts. Friends of James River Park, James River Outdoor Coalition, the Richmond affiliate of the Mid-Atlantic Off-Road Enthusiasts and many other organizations have played a critical role in the advancement and maintenance of the park.

The conservation easement, in addition to protecting riverside properties from sale or development, signifies city government recognition of the value of the river and the park. The recently completed master plan maintains the direction the park has taken over its history—keep it natural and open to visitors. The plan also recognizes the tension generated by those two principles. Is it possible to accommodate the increasing number of visitors and maintain the wild and natural character that is so highly valued? Planners, managers, advocates and visitors hope so.

———

The cleaner river and the establishment of the park had made great strides toward wildlife habitat, but one historic feature remained an obstacle—dams.

5

DAMS, FISH, EAGLES, GEESE AND MORE

A thing is right when it tends to preserve the integrity, stability and beauty of the biotic community. It is wrong when it tends otherwise.
—*Aldo Leopold,* A Sand County Almanac

Without habitat, there is no wildlife. It's that simple.
—*Wildlife Habitat Canada*

Cleaning up the James River and maintaining natural habitat in the park have resulted in a flourishing variety of wildlife in and along the river. A list includes osprey, bald eagles, cormorants, ducks, Canada geese, great blue herons, falcons, hawks, turkeys and even an occasional pair of swans; muskrats, beaver, deer, mink, river otter, fox, coyote, groundhog, opossum, raccoon, squirrel and, the favorite of some, salamander. (See Science in the Park on the James River Park website.) The cleaner river made for healthier fish populations too, but a problem remained. Dams prevented anadromous fish (that spawn in freshwater but spend most of their life at sea) from returning to their historic spawning grounds. The number of American shad making their annual spring run on the James had declined precipitously. Of course, dams were not the only cause of the decline; pollution and overfishing played their part.

FISH PASSAGES: BREACHING DAMS WITH A LITTLE HELP FROM NATURE

In an earlier time, building dams was a sign of advancement; in the latter part of the twentieth century, removing dams became the mark of progress. The environmental movement in Richmond and the rest of the country drew attention to the detrimental impact of dams. So, the task of removing dams, or rather modifying them so shad and other migratory fish could pass, became part of the river agenda in Richmond. The Virginia General Assembly, the City of Richmond, the Virginia Department of Game and Inland Fisheries and a number of volunteer and advocacy groups were involved. An early action stemming from that issue in Virginia was House Joint Resolution No. 233 passed by the Virginia General Assembly in 1981. Citing the concern that "anadromous fishes are prevented by low profile dams at Richmond, Virginia, from moving upstream to their historical spawning and rearing areas," the resolution called for state and local institutions to determine the need "as may exist for the construction and maintenance of devices to pass fish along the James River…and to take such action as may be feasible and effective for providing access for anadromous fishes to their historical spawning and rearing sites."[89] Thus began a twenty-year series of meetings, arguments, studies, plans and fundraising that resulted in opening a path to traditional spawning locations.

THE FIVE DAMS

The James River in Richmond has five dams originally built for water supply, power generation, watering canals and manufacturing. For well over a century, these dams prevented anadromous fish from reaching their traditional spawning grounds.

The first dam encountered by migratory fish swimming upstream is the Manchester Dam zigzagging across the river within one hundred yards or so of the tidal James. Constructed to divert water to the Manchester Canal, the water was used for milling operations, hydropower and water supply on the south side of the river. Begun as a wing dam about 1803, it was extended in 1858 and again in 1878 to cross the full width of the river.

Brown's Island Dam is immediately upstream of the Manchester Dam and nearly joins it on the north side of the river. It was constructed in 1901

Moving upstream, Manchester Dam is the first of five dams that blocked spawning runs. Brown's Island Dam (now T. Tyler Potterfield Memorial Bridge) is just upstream. *Photo by author.*

to divert water to the 12th Street Hydroplant and a steam electric plant owned by Virginia Electric and Power Company (VEPCO, now Dominion Energy). This dam consists of multiple gated openings that could be raised or lowered to control water flow.[90]

The lowhead dam on the north side of Williams Island. *Photo by author.*

Not far upstream is a series of three connecting dams that essentially form only one: Hollywood Dam, North Belle Isle Dam and Belle Isle Dam.[91] The first of these, Hollywood on the north side, was built in 1830 to supply water for the city. It was later rebuilt to divert water to the Hollywood Hydroplant. North Belle Isle Dam continues the structure to the upstream tip of Belle Isle. Belle Isle Dam runs from that same upstream tip of Belle Isle to the south shore and was designed to funnel water to a hydroplant on the south side of Belle Isle.

Several miles upstream, the next obstruction is Williams Dam. Williams is actually two structures, one on the north and one on the south of Williams Island. This dam diverts water into an intake canal leading to the City of Richmond's water treatment facility. The northern dam segment was constructed in 1905, and the southern (also called Z Dam) was built in 1932 to replace an earlier loose stone dam.[92] The southern Z Dam keeps water flowing into the channel north of the island and then into the intake canal.

Bosher's Dam, approximately three miles upstream from Williams Island, is the last of the five Richmond dams and, at roughly ten feet, is the tallest of the five. It was built in 1823 and extensively rebuilt in 1835. Bosher's Dam was constructed by the James River and Kanawha Canal Company to divert water to the navigation canal along the falls to downtown Richmond. It was later used to supply water to Byrd Park Hydroplant and the Hollywood Hydroplant.[93] It still serves as a backup source of water for the city water treatment facility during periods of low flow, and, not incidentally from a political viewpoint, it creates a lake with a shoreline dotted with expensive homes and a playground for powerboaters.

THE FIRST BREACHES

The 1981 Resolution in the General Assembly resulted in the *Feasibility Study of Fish Passage Facilities in the James River, Richmond, Virginia*.[94] This report proposed fish passage facilities, for both upstream and downstream travel, for the five Richmond dams. After the feasibility study and VEPCO's decision not to restore hydropower operations, there was little disagreement

First break rapid. The Hurricane Camille flood in 1969 punched a hole in Hollywood Dam that has continued to expand—a favorite spot for Passages Camp paddling instruction. *Photo by author.*

about the desirability of breaching Manchester and Brown's Island Dams. The rub was funding. For a time, the matter seemed at stalemate, but the General Assembly came to the rescue with the creation of the Virginia Fish Passage and Revolving Fund sponsored by Virginia senator Joseph B. Benedetti. This fund would contribute to payment for fish passages on the James and other rivers in Virginia and began a public-private partnership that supported the breaches in the five Richmond dams.

Early in 1989, breaches were blasted in both the Manchester and Brown's Island Dams. Since a natural breach already existed at Belle Isle, a result of the Hurricane Camille flood, passageway was now open for an additional six miles to Williams Dam.

ON UPSTREAM

Even before funding was secured, serious discussion and planning for Williams Dam got underway. One catalyst for action on Williams was a drowning death in the hydraulic at the base of the dam, bringing the number of known deaths there to sixteen. The Falls of the James Scenic River Advisory Committee began to press for a solution that simultaneously would make the dam safer for paddlers and allow fish passage.[95] This dual objective became the focus of discussion among the Richmond Department of Public Utilities, Richmond Planning Department, James River Park System, Virginia Department of Conservation and Recreation, Virginia Commonwealth University, Virginia Department of Game and Inland Fisheries and the U.S. Fish and Wildlife Service. The central idea was to cut a notch in the dam at a depth that would allow fish passage but would not lower the water level behind the dam enough to affect flow into the water intake canal on the north side of Williams Island.[96] Two studies were prerequisite to any go-ahead: a water flow analysis to determine the drop in water level that would result from the notch and a study to ascertain the structural integrity of the dam to ensure that it could withstand the proposed cut. It was concluded that the water level drop would be negligible, and in any event, "the impact will be acceptable given the city's ability to utilize the Kanawha Canal during low River flows."[97] It was determined that the dam was structurally capable of withstanding the proposed cut. The plan, still awaiting funding, seemed to be in place.

In anticipation of a passageway being established, a second leg of the strategy was initiated: restocking upstream. The idea was to hatch American shad fry and place them above all the dams in the hope that the upstream location would be imprinted and lead to their return as adults after passageways were opened. Brood stock from the Pamunkey River and to a lesser extent the Potomac were used at the U.S. Fish and Wildlife Service Harrison Lake National Fish Hatchery to produce the fry. Their youthful passage downstream over the dams was considered no problem.

The Lower James River Association (later to drop the "lower" and become the James River Association) under the leadership of Patricia A. "Patti" Jackson launched a fundraising campaign in 1992 to fund both the Williams and Bosher's projects. The estimated cost, to go up later, was $950,000.[98] Others, like the Virginia Anglers Club, the Virginia Bass Federation and a number of corporations, pitched in through donations, fundraising projects and grants.[99]

In the fall of 1993, the notch in Williams Dam was cut, two and a half feet in depth, thirty feet wide and ninety feet from the south shore.

The notch in Williams (Z) Dam was intended to be good for fish and safe for paddlers but found to be potentially dangerous for paddlers at some water levels. *Photo by author.*

And so, a new issue was raised and dramatically framed by Charles Ware in the title of a communication to river groups: "WILLIAMS ISLAND DAM NOW SAFE FOR FISH—BUT UNSAFE FOR BOATERS" (March 1994). The second objective, making the notch a safe passage for paddlers, was not achieved, at least at some water levels. The Falls of the James Scenic River Advisory Committee and Coastal Canoeists, among others, recommended grout bags or another engineering solution as a possible retrofit to make the notch paddler safe, but no action has been taken. Indeed, some kayakers now resist any alteration since the notch produces a desirable surfing wave below the break that grout bags might eliminate. The solution to the problem created, or not solved, by the notch as constructed is a warning sign encouraging boaters not to paddle through the notch.

ON TO THE "INSUPERABLE BARRIER"—BOSHER'S DAM

Creating passage in Bosher's Dam, called an "insuperable barrier" in one report, was a larger task than Williams, both physically and financially.[100] Unlike Williams Dam, a simple breach was not possible, nor—unlike Embrey Dam on the Rappahannock near Fredericksburg—was complete removal, since this dam serves to divert water into the old James River and Kanawha Canal that serves as a backup supply at low water level for the city water treatment plant. An additional hurdle was ownership. CSX, not the city, held title to the dam, even though, through agreement, the city managed dam operations. The Bosher's Dam vertical slot fishway took six years to complete rather than the two years originally thought because of delays in the city's negotiations with CSX about ownership, because bids came in higher than expected, requiring more time to raise additional funds, and because of weather.[101]

The groundbreaking took place on July 22, 1997, and with the bulk of summer still ahead, the hope then was for fishway completion in time for the spring 1998 spawning run. But that was not to be. Leaking coffer dams caused delays in the summer and fall of 1997, and floods and wet weather continued the delays for months beyond. Persistence paid off, however. The fishway was completed in February 1999, in time for that year's spring spawning season, and was dedicated on April 20, 1999. According to Alan Weaver, fish passages coordinator for the Virginia Department of Game

American shad is the target fish for the vertical slot fishway at Bosher's Dam. Shad numbers have been small, but the fishway is used by large numbers of other species. *Photo by author.*

and Inland Fisheries, the final cost was $1.6 million, with approximately 75 percent from government sources (federal, state and local).[102] The remainder, primarily from corporations and private foundations, was raised by the James River Association and others over a five-year period.[103]

THE NEXT QUESTION: WOULD IT WORK?

The effort to open the five dams in Richmond took twenty years, or more, depending on the starting point one chooses. From the first physical act, blasting the first two dams on the path upstream, ten years passed before the final opening was achieved. The projected payoff was to be a healthier river, returning at least a bit of its historic character, as well as robust economic returns from fishing. The answer to the key question was up to the fish: would they use the passages and return in the numbers hoped for?

Optimism was high and seemed warranted; nearly 800 American shad swam through the Bosher's Dam fishway in 2002. That seemed to bode well for the future, building the expectation for thousands in the years to follow. But the next several years were thoroughly disappointing. In 2003, fewer than 200 fish passed through and only about 50, an all-time low, in 2008.[104] Still, the restoration program, which began in 1992, continued until 2017. As of 2017, 125,846,446 shad fry had been stocked in the upper James. That year, the Department of Game and Inland Fisheries (DGIF) seemed to throw in the towel. Despite carrying out an aggressive stocking program for more than two decades and breaching in one fashion or another all of the downstream dams, results were far less than expected or hoped, so "DGIF will not be stocking American Shad in the foreseeable future."[105]

Why the American shad population has not recovered as hoped is still not fully explained. Are the passages inadequate, and would complete dam removal have made a difference? Are predator fish reducing shad numbers? Is the imprinting theory somehow flawed? Is it the impact of offshore commercial overfishing? Perhaps Dr. Greg Garman's conjecture is accurate; human activity may push a species "to a biological point of no return, where no amount of time or effort can turn things around."[106]

OTHER FISH

Atlantic Sturgeon

A nascent recovery that has spurred a great deal of excitement is the resurgence of "fish that swam with the dinosaurs"—the Atlantic sturgeon. Although they had survived for seventy million years, Atlantic sturgeon were thought to be on the edge of extinction before what now appears to be a comeback. An increasing number of sturgeon have been sighted in the James River and elsewhere.

A subject of legends—Indian boys riding them as a rite of passage, a food source that enabled the settlers at Jamestown to survive—sturgeon faced severe decline as a result of overfishing, dams, pollution, sedimentation and dredging that restricted spawning areas. In addition to habitat deterioration, several features have made sturgeon vulnerable. Historically, the value placed on their roe, used for caviar, made them a sought-after target. A moratorium on sturgeon fishing has eliminated that threat on the James,

Jumping sturgeon, a rare sight, photographed just downstream from Richmond. *Courtesy of Don West.*

although not in other parts of the world. Additionally, sturgeon are slow to reproduce. They take close to two decades to reach sexual maturity and do not reproduce every year. Sturgeon also require higher levels of oxygen, so pollution is especially destructive. In addition to the fishing moratorium, the improvement in the quality of the river environment may be the essential ingredient in their comeback.

For more than a decade now, research and restoration on the James have been underway, sponsored and conducted by a consortium of federal agencies, universities and advocacy groups: U.S. Fish and Wildlife Service, National Oceanic and Atmospheric Administration, Virginia Institute of Marine Sciences, Virginia Commonwealth University (VCU) Rice Rivers Center, U.S. Army Corps of Engineers, National Fish and Wildlife Foundation and the James River Association.[107] So, Atlantic sturgeon in the James River have quite an array of organizations working on their behalf.

One aspect of the restoration effort is the construction of spawning reefs—football field–size areas covered with a two-foot depth of granite donated by Luck Stone.[108] Channel dredging and sedimentation has drastically reduced the hard, clean bottoms sturgeon need for spawning, so the reefs are intended to replace some of that habitat and provide hiding places for young sturgeon during their early months.

The VCU sturgeon research team is focused on the movement and reproductive behavior of the Atlantic sturgeon in the James River. The team's on-the-water leader, Dr. Matt Balazik, has caught and tagged well over one hundred adult sturgeon. Tagging with electronic monitors allows

Constructing a reef of granite to provide spawning habitat for sturgeon. *Courtesy of James River Association.*

sturgeon to be tracked over time and movement patterns analyzed. Some have been detected within the city limits near Ancarrow's Landing and in the Mayo's Bridge area. With the James River in an improving condition and with help from its devoted friends, the future of sturgeon on the James looks brighter than it has for a century.

Catfish: Boon or Bio-Pollution

While the vitality of American shad recovery is disappointing and the restoration of sturgeon is hopeful but uncertain, the nonnative catfish population has been on an upward trajectory. The introduction of nonnative catfish in the James River has produced highly popular fishing opportunities, a reduction, at least arguably, in other fisheries and a concern about the health of the ecosystem. "Giant catfish are taking over the James River." So began an article in the April 27, 2008 edition of the *Richmond Times-Dispatch*. Simultaneous with this concern about takeover, these same fish are celebrated with picture after picture in newspaper articles and the websites of fishing guides. Are catfish an invasive species that threatens the ecological health of the river or a desirable gamefish that simply makes the river different but still environmentally sound? That argument has been simmering for some time.

Blue catfish, native to the Mississippi watershed, were introduced in the James River in the 1970s to expand game-fishing opportunities. And indeed they are a popular quarry for anglers, both local and from out-of-state. The blues have adapted well to the new environment, multiplying rapidly and growing to a substantial size—approaching 100 pounds in some cases. The current record catch, caught in 2009 in the tidal James, is 102 pounds. Flatheads, too, are nonnative, having been introduced in the James in the late 1960s.[109] Widespread in the bay watershed, but with a smaller range than blue catfish, they are a significant part of the fishery in the Richmond region.

The issue is far broader than just Richmond, encompassing the Chesapeake Bay watershed and several states, but the James in Richmond is in the mix since it contains both blues and flatheads and is considered a prime bass fishery as well as the gateway to traditional shad spawning

This flathead catfish was caught above the fall line by a young fisherman with the help of Mike Ostrander's Discover the James guide service. *Photo by author.*

According to legend, Richmond was named because of this view of the James River, which resembled the view of the River Thames at Richmond-upon-Thames in England. *Photo by author.*

RICHMOND.
FROM THE HILL ABOVE THE WATERWORKS.

This classic painting suggests the view as it might have been in an earlier time from the hill that is now Hollywood Cemetery. *Courtesy of the Library of Congress.*

The view from Hollywood Cemetery as it is today. *Photo by author.*

The downtown Richmond skyline seen from the newly constructed T. Tyler Potterfield Memorial Bridge. *Photo by author.*

Rail has been and continues to be an important part of Richmond's riverfront. *"Richmond's James River View" by Adam N. Goldsmith, courtesy of Scenic Virginia.*

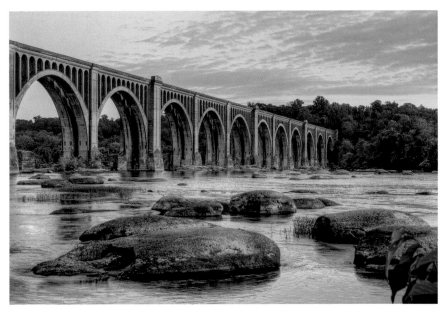

The graceful arches of the Belt Line Bridge make it one of the most photographed scenes along the river. *"Railroad Bridge at Sunset" by Bill Piper, courtesy of Scenic Virginia.*

Day dawns over Pony Pasture Rapids. *Courtesy of Scott Adams.*

Hollywood, the James River's most celebrated rapid. *Photo by author.*

A tranquil scene along the river. *Photo by author.*

A view of Rattlesnake Creek, renowned for having no rattlesnakes. *Photo by author.*

The James River in Richmond is a bountiful subject for photographic artistry, no matter the season. *Courtesy of Scott Adams.*

Water artistry. *Courtesy of Scott Weaver.*

A chilly scene with great beauty. *"Winter Tree on James River" by Harold Lanna, courtesy of Scenic Virginia.*

Paddleboarders on a foggy morning. *Courtesy of Tricia Pearsall.*

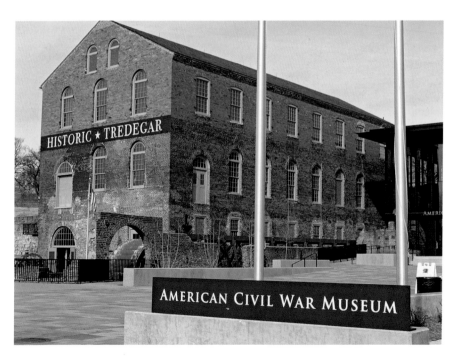

Two Civil War museums provide an anchor for the showcase of history along the riverfront. *Photo by author.*

The Great Ship Lock was the final connection between the Kanawha Canal and the James River. It could be returned to working order. *Photo by author.*

The Headman by Paul DiPasquale commemorates the role of African Americans on the James River. *Photo by author.*

Left: The Christopher Newport Cross, relocated to the Canal Walk in 2003, commemorates Englishmen's first arrival at the Falls of the James in 1607. *Photo by author.*

Below: The life-size statue of Lincoln and his son Tad behind the Tredegar site of the Richmond National Battlefield Park memorializes their visit after the fall of Richmond in April 1865. *Photo by author.*

Byrd Park Pump House, an example of Victorian Gothic architecture, is now part of the James River Park System. *Photo by author.*

The Canal Walk is a signature project in the transformation of the downtown riverfront in Richmond. *Photo by author.*

The recently completed and instantly popular T. Tyler Potterfield Memorial Bridge is the first major project of the 2012 Riverfront Plan. *Photo by author.*

Canada geese are a well-entrenched species along the James River in Richmond. They can be a beautiful sight to see and a decided nuisance. *"Pony Pasture Geese" by Scott Adams, courtesy of Scenic Virginia.*

Great blue herons, once rare, offer an elegant sight at many locations along the James. *Courtesy of Scott Weaver.*

Eagles have made an incredible comeback along the James River, including in the city of Richmond. *"Eagle on the James River" by Edward Episcopo, courtesy of Scenic Virginia.*

Osprey return to the James in the spring to raise their young. *"Osprey Waiting for Dad's Fish"* *by Edward Episcopo, courtesy of Scenic Virginia.*

This notch in Z-Dam was meant to attract migrating fish, but sometimes paddlers can't resist. *Courtesy of Scott Adams.*

Some paddlers like it best at flood stage. *Photo by author.*

Runners take advantage of the miles and miles of trails along and across the river. *Photo by author.*

Just another summer day on the river for dog and owner. *Courtesy of Scott Adams.*

Sometimes the river is just for relaxing. *Photo by author.*

grounds. Blue catfish are more numerous in the tidal portion of the James, downstream from the 14th Street Bridge. Upstream, flatheads are more common and a favorite target of fishermen.

In 2012, the Chesapeake Bay's Sustainable Fisheries Goal Implementation Team Executive Committee issued a statement proclaiming that it "has concluded that the potential risk posed by blue catfish and flathead catfish on native species warrants action to examine potential measures to reduce the densities and limit range expansion, and to evaluate possible negative ecological impacts."[110] Similarly, the National Oceanic and Atmospheric Administration (NOAA) has declared these nonnative catfish as "invasive."

> *An invasive species is defined as an "alien species whose introduction does or is likely to cause economic or environmental harm to human health" (Executive Order 13112).*
>
> *Blue and flathead catfish are considered invasive species in the Chesapeake Bay....Blue catfish comprise a highly valued recreational fishery in some areas, but both blue and flathead catfish are likely negatively affecting native species and the Chesapeake Bay ecosystem.*[111]

Some anglers and fishing guides believe the "invasive" label gives catfish a reputation that is not deserved.

To what extent are these imported catfish culprits in damaging other fish populations, including native species? If the concern is the effect of blue and flathead catfish predation on other populations, especially smallmouth bass (also nonnative), bream and shad, it is important to know what these catfish eat. If both are apex predators, eating at the top of the food chain, and their numbers and biomass are increasing, the worry would seem justified that the imported catfish are endangering native species, including the shad that have been subsidized by millions of dollars. Or is it the case that the predominant food source for flatheads is bottom-feeding suckers, not bass, shad and native catfish? Biologists (and fishermen) have been working to sort out the evidence, both scientific and anecdotal: what do catfish eat? Studies are underway to determine more precisely the dietary habits of both flatheads and blues. Evidence is cumulating but no final answers. One point is without dispute: even if catfish are a contributing factor in the failed restoration of American shad on the James, it is only one of a number of factors.

OUT OF THE WATER WILDLIFE

Geese

Geese are now a year-round fixture on the James in Richmond and considered a problem, just as they are in parks, airports, golf courses, farmers' fields and even shopping malls around the country. The iconic image of Canada geese is the V-shaped skein flying south for the winter or north for the summer. The problematic geese are not the awe-inspiring formation heading north or south, but nonmigratory year-round residents. They are "resident Canada geese," defined by the U.S. Fish and Wildlife Service as those "nesting within the coterminous United States in the months of March, April, May, or June, or residing within the coterminous United States in the months of April, May, June, July, or August."[112]

Human intervention is the reason there are so many resident, not migratory, geese throughout the country and on the James in Richmond. The population of resident Canada geese is "the result of purposeful introductions by management agencies, coupled with released birds from private aviculturists and releases from captive decoy flocks after live decoys were outlawed for hunting in the 1930s."[113]

Resident Canada geese are a beautiful sight on the river, whether gliding through flat water, floating through a rapid, leading a troupe of goslings or

Sometimes Canada geese can be downright rude. *Photo by author.*

Bushnell ⓜ 55F13C ● 02-09-2018 13:52:10

A coyote "caught" by a Science in the Park game camera. *Courtesy of Anne Wright.*

coming in for a water landing. So why are they considered a problem? A large part of the answer is captured in a *Richmond Times-Dispatch* headline: "For Canada Geese, River Is Their Toilet."[114] Their numbers have grown so large that during long periods of moderate to low water levels, almost every rock in the river seems to be covered with "goose poop."

In small numbers, geese add a sense of natural beauty to the river setting; in large numbers, they make the river experience seem like hiking or picnicking in an overstocked barnyard. Sporadic efforts have been made along the river to control geese numbers. The James River Park has led some efforts at control through oiling or addling eggs in the nest, but these efforts have not been made with the intensity or consistency required to reduce the numbers to a level that avoids the problems. Their numbers remain large. Hunting is not allowed within the city limits, Canada geese are protected by the Migratory Bird Treaty Act and resident geese have a longer breeding season than migratory geese. Resident geese have faced no natural predators in Richmond, but that is changing with the return of coyotes. Coyote presence is documented by Anne Wright's Science in the Park game cameras. Coyotes eat eggs, goslings and adult geese.

Eagles

The resurgence of bald eagles on the James River is a clear sign of the environmental recovery that has occurred. Although most of the James River eagles call the area downstream from Richmond home, a few do nest within the city limits and eagles can be seen with some frequency in the city. In 2012, an "eagle-cam" was installed above the nest of a pair within the city by the Center for Conservation Biology and the *Richmond Times-Dispatch*.[115]

By the mid-1970s, there were no pairs of eagles along the James; now the James River watershed is home to more than 250 pairs. Banning DDT and the recovery from the Kepone spill have been key to eagle population growth. Although still protected, bald eagles were removed from the federal list of endangered and threatened species in 2007 and taken off the Virginia list in 2013. While the bald eagle comeback is a national phenomenon, the tidal portion of the James has led the way.

Eagles have even become a tourism and recreation draw with the initiation of bald eagle tours by Discover the James headed by Mike Ostrander—Captain Mike, as he is called on the water. Captain Mike leads regular eagle-sighting tours and has given names to several of the resident eagles he sees frequently: Bandit, Varina, Enon, Duke, Baba, Pops.

Great Blue Herons

Great blue herons are another sign of an expanding wildlife population resulting from a healthier environment. While great blues can be seen as solitary figures feeding along the river's edge, the excitement about their place in Richmond focused for a time on the large heron rookery on Vauxhall Island, just upstream of 14th Street Bridge (Mayo's), where the rapids meet the tidal section. Beginning about 2007, herons were seen hatching chicks in large nests atop trees on the island. About forty herons made this their nesting ground. Unfortunately, the herons did not return to this downtown Richmond location in 2015; perhaps they will one day.

Although herons hunt and feed as singles, they nest in groups. Pairs are monogamous during a season but choose a new mate each year. Both parents share in nest building, sitting on eggs and feeding chicks after they hatch. Chicks are hatched in spring when the shad run and food is plentiful.

Eagles have made a dramatic comeback along the James River, including in Richmond. *Courtesy of Tricia Pearsall.*

For eight years, this great blue heron rookery in downtown Richmond was a wildlife fan favorite, but the herons did not return in 2015. Perhaps they sought more privacy. *Courtesy of Scott Weaver.*

Spotted Salamanders

Spotted salamanders have become an iconic symbol signifying that "the urban river is still wild."[116] The Friends of James River Park adopted the spotted salamander as its mascot and features it on cards and automobile stickers. These six-to-seven-inch amphibians live largely hidden for most of the year and emerge on rainy wet evenings in late February or early March to mate in small vernal pools. These shallow pools dry up in summer and so do not hold fish that would eat the young as they hatch from eggs. It has become a local tradition to track down and observe the salamanders during the period when they venture out to the mating pools. Riverside Drive between Pony Pasture Rapids and Huguenot Flatwater is the favorite viewing site.

An unsuspecting riparian owner along Riverside Drive discovered the passionate regard some hold for salamanders after he cleared and planted grass on the front of his property, which slopes toward the river. River advocates complained that the clearing removed protective salamander habitat. They made the issue public and the confrontation became a featured story in the local newspaper.[117]

Unlike bald eagles, however, the population of salamanders may be on the decline, the result of lost habitat as more homes, trails and parking lots line the James. A decline in salamanders has been observed, but verifiable numbers are hard to come by. Are the numbers declining in the entire Richmond reach or just in the most viewed area along Riverside Drive where cars and brush-clearing homeowners are the enemies of this cute little amphibian? Anne Wright reports finding vernal pools containing salamander egg masses in other areas of the park, so perhaps the salamander population is more vibrant than sometimes thought.

OTHER WILDLIFE

Many more species of land and water wildlife make the James River in Richmond home. Maymont Nature Center focuses on James River fish and wildlife and is an excellent place to see a variety of species up close. Although no claims can be made that pristine conditions have been achieved, the river and its banks provide a welcoming habitat. Many species are seldom visible and certainly not given the notoriety of eagles, great blue herons

River otters are back, as evidenced by this Science in the Park game camera photo. *Courtesy of Anne Wright.*

or even salamanders. Although seldom seen, river otters are back. Coyotes too have returned and suggest a more "balanced nature" that may keep the resident goose population in check. A birdwatcher no doubt could list dozens of species found along the river. And of course, there are snakes, mostly harmless, and insects. Ralph White can describe in poetic terms the types of fireflies found on a summer evening in The Wetlands Park.

PERSPECTIVE

The transformation of the James River has been a boost to wildlife on the James in Richmond and beyond. Eagles, osprey and cormorants are flying, and the fishing is exceptional—although what is caught may be different than it would have been without human intervention. Sturgeon can be seen breaching in September and October in the tidal waters, and salamanders still cross Riverside Drive on wet winter nights. The return of mink and river otters adds to the environmental good news. These species need clean conditions and abundant food to survive, so their presence is confirmation of an improving ecosystem.

The heron rookery in downtown Richmond was greeted with great fanfare when it began in 2007, but disappointment followed when herons did not return to the rookery in 2015. This may simply be a natural migration, but it reduces the human entertainment value of the herons. Despite the absence of the rookery in a prominent downtown location, great blue herons are still a common and majestic sight along the shoreline of the river. Of course, wildlife often surprises us. Even a manatee has been spotted at the base of the fall line in Richmond. The manatee may have lost his or her internal GPS for a time, or be a product of global warming, but it no doubt would have been turned back by the wall of pollution it would have encountered in the 1950s.

In addition to surprises, there are questions. Why have American shad not returned upstream to spawn in the numbers expected? Is the population of spotted salamanders stable, expanding or on the decline? What is the significance of the proliferation of nonnative catfish? Will the return of coyotes to the banks of the James help keep the Canada geese population in check? Can wildlife habitat be sustained in the face of the growing number of humans who visit the James River Park? Apart from these specific issues, we can also ask the broader question: by what standard are we to judge the health of a modern ecosystem?

One mental model of a healthy environment and wildlife habitat is to approximate as closely as possible the state of nature before human intervention. By that standard, ultimate success would be a river with a robust population of American shad and other anadromous fish returning to their historic spawning habitat and nonnative catfish (and humans?) eliminated. That standard, of course, is unreachable. So, would a clean river with a healthy and robust catfish population, an ample supply of resident Canada geese and English ivy on the trees be a satisfactory replacement?

———

Wildlife is the basis for much of the recreation along the James River in Richmond: birdwatching, fishing, salamander scouting—so are swimming, paddling, hiking, climbing and more. Creating the conditions for all that fun has brought out its share of conflict as well as cooperation and passion.

6

RECREATION

A LITTLE CONFLICT MIXED WITH THE FUN

There is a little Huck Finn in all of us.
—*Daniel McCool*, River Republic

*The Surprising Science That Shows How Being Near, In, On, or Under Water
Can Make You Happier, Healthier, More Connected, and Better at What You Do*
—*subtitle of* Blue Mind *by Wallace J. Nichols*

The recreational use of the James River in Richmond has grown
dramatically. Many indicators reflect the current recreational
popularity of the river: the nearly two million visits to the James
River Park each year, the thousands who have taken a commercial raft
ride through downtown, the tourist rides on the downtown canalboats, the
overflow parking at Huguenot Flatwater and Pony Pasture Rapids on any
summer weekend, the many fishing boats in the tidal section on a spring
day when the shad are running, the numerous visitors seeking a peek at
history along the river, the concerts and festivals held on Brown's Island
and other riverside venues, the number of articles in the local press about
recreational opportunities along the river, the variety of activities listed in
the James River Advisory Council's *James River Days*, Richmond's selection
in 2012 by *Outside Magazine* as "the best river town in America" and more.

The James River Park System is the centerpiece of recreation on and
along the river as former councilwoman Kathy Graziano's newsletter brags:

The park's wilderness shoreline amid woods, islands, meadows and rocks offers opportunities for canoeing, kayaking, tubing, walking, jogging, hiking, rock-climbing, biking, fishing, boating, swimming, sunning, bird-watching, and nature study, with many of these activities just blocks from the City's downtown residential, business and financial district.[118]

The park is the centerpiece but only a part of the recreational opportunities the river now affords. Despite the limited access in the past, the river has been a draw for recreation for decades—even centuries. Note, for example, the use of Pony Pasture before the park was established and the recreational use of Vauxall Island in the 1800s. On Vauxall, "the Quoit Club, composed of the notable gentlemen of Richmond celebrated there.…There they had their outdoor sports and exercised their comraderie. Many a bowl of apple toddy, and many a glass of Julip aided and abetted in festivities of this club."[119]

The river as it is, however, does not please everyone. A recent park visitor described his experience on TripAdvisor as "a bad time in the wilderness. No amenities, no shelters, no restaurants." Of course, some river and park enthusiasts would likely say, "Exactly, just like we want it." It is true that a visitor cannot show up at the river with no preparation and expect to be entertained Disney-style. At least much of it, certainly the park, is a make-your-own-entertainment kind of place, unless you make arrangements with one of the outfitters.

PADDLING

Topography, as noted in the introductory chapter and indicated on the map, is one feature that makes the James in Richmond such a draw for paddlers: there is something for everyone. In the ten or so miles from Bosher's Dam on the west to Rocketts Landing on the east, the elevation drops more than one hundred feet. The elevation drop is not continuous, however, providing distinctive sections that accommodate different recreational uses and drawing visitors from miles away. For boaters, the river in Richmond has four distinct sections.

The area from Bosher's Dam to Z Dam is flat with slow current (except in high water) made accessible by Huguenot Flatwater, a section of the James River Park System. Recreational kayaks, canoes and stand-up paddleboards frequent this reach of the river, especially on summer evenings and weekends. From Z Dam to the western tip of Belle Isle, the periodic rapids pick up intensity to Class II whitewater, so some paddling skill is required if the chance for an unexpected swim is to be avoided. Access for launching is provided by the James River Park at Pony Pasture Rapids, the area's most popular swimming hole, and takeout is at Reedy Creek in the main section of the park. Reedy Creek is also the launch site for the more challenging "downtown" run. Between the western tip of Belle Isle and the tidewater portion of the river marked by Mayo's Bridge, the elevation drop increases to seventy feet over two miles and the several rapids increase in intensity to Class III or IV depending on water level. From Mayo's Bridge downstream, the water is flat, slow-moving and deep, a place more accommodating for motorized craft, sea kayaks, racing shells and even dragon boats on occasion.

So, this short reach of river from Bosher's Dam to Rocketts Landing provides calm water paddling, water requiring intermediate whitewater skills and a section for more advanced paddlers, capped by the tidal section for motor craft, shells or touring kayaks.

CANOE/KAYAK/PADDLEBOARD

Paddling on the James, like other waters, has grown almost exponentially since the time in 1975 when a paddler received a letter from the James River Park superintendent warning him, "At this time I would like to advise you that boat launching is not permitted in the James River Park. Violaters [sic] are subject to arrest by Park Rangers and City Police." That boating prohibition policy changed soon thereafter, and paddling popularity was on the rise, supported by manufacturers providing a seemingly never-ending variety of canoes, kayaks and paddleboards of almost every imaginable shape and size. Instructional and guide support have grown apace.

Paddling classes are readily available from both public and private sources, including Black Dog Paddle, Chesterfield County Parks and Recreation, Riverside Outfitters, River City Adventures and RVA Paddle Sports. Additionally, the James River Park now offers on-river instruction. Chesterfield County has been an area leader in outdoor instruction for many years. Greg Velzy and a stable of certified instructors offer training in multiple forms of kayaking as well as canoeing and stand-up

Paddleboarders making their way down Pipeline Rapids. *Courtesy of rich young.*

paddleboarding. Beyond Boundaries offers trips and a variety of outdoor experiences for persons with disabilities.

Stand-up paddleboarding (SUP) has come on the river scene like gangbusters in the past few years. Paddleboarders, some struggling to stay upright and others as smooth as glass, are now a common sight on flatwater, but stand-up paddlers by no means stick to flatwater. They can be seen going through the best whitewater Richmond has to offer. It seems that dog-paddling also is popular—dogs as passengers, that is, on canoes, kayaks and paddleboards. Some of these canine passengers are even decked out with lifejackets. Some enjoy an occasional swim, while others work very hard to stay dry.

Paddlers not only use the river but also have been active in a variety of ways in improving the paddling environment. When the city threatened to ban boating on the river in high water, paddlers themselves led the formulation and initiation of the "5/9" rule. Under this guideline, a PFD (life vest) is required when the water is above five feet at the Westham gauge, and a permit is required to boat when the river is above nine feet. Below five feet, there are no restrictions, although the James River Park attempts through signage and education to inform users of the inherent risks such as drop-offs when wading. Paddlers (as well as trail enthusiasts and climbers) have contributed many thousands of hours and dollars over the years, especially since the James River Outdoor Coalition (JROC) was formed in 1998. In 2018, JROC contributed $20,000 through its 20 for 20 Project on the occasion of its twentieth anniversary.

Paddling competitions have been a part of the James River scene at least since 1971 when the 1st Annual Belle Isle Whitewater Slalom was held on the south side of the river adjacent to Belle Isle. That race, despite its title, did not meet the aspirations of its organizers to become an annual event. Other competitions, however, have been held over the years, including the Urban Whitewater Race in 1983 and the Open Canoe National Championships in 1998.

At least on one occasion, the prospect of paddling competition was a source of conflict. On a Wednesday evening in the late winter of 1999, seventy or so whitewater boaters met in the Reedy Creek Park Headquarters. The discussion was passionate but mostly civil. The issue was whether to invite ESPN to film a whitewater rodeo (competition involving kayak and canoe tricks in a rapid) as part of its X-Games trial series. The hope on the part of those who proposed the idea was that a similar event would be held annually thereafter on the James River.

A "rodeo move" in downtown Richmond. *Courtesy of rich_young.*

Some, however, argued that the televised event and annual competitions would bring too many boaters to the river and ruin the ambience for those who prefer a more natural environment. This kind of event and the crowds it might attract would threaten what makes the river special, that feeling of being in a natural setting, even in the center of the city. Proponents argued that the event would help bring attention to the importance of the river, making it more likely that the river would be protected as a natural recreation resource. Further, it would help promote the very sport that brought the group together.

The emotionally charged discussion led to a show of hands, defeating the proposal by one vote. A request for a recount to assure that the original tally had been accurate was not allowed by the winning side on the ground that it was an attempt to manipulate the outcome. Passions ran so high over the issue that friendships reportedly were lost, or at least weakened.[120]

This meeting in itself could not be classified as one of the significant events shaping the use of the James River. The boaters assembled did not have authority to determine whether a whitewater rodeo was held or even whether an invitation was extended, but the conflict does show the passion that some feel toward the river; caring for the river can be controversial. At times, commercial rafting also stirred conflict along with the enjoyment it has provided over several decades.

COMMERCIAL RAFTING

The iconic image of Richmond as a "whitewater city" is the picture of a raft of barely visible customers and a smiling guide blasting through a wall of whitewater. This photo, better than words, captures the spirit of recreation on this urban river. A trip with a rafting outfitter is an easy way for someone to experience the river without investing in equipment or paddling skill.

Commercial rafting in Richmond was launched in the mid-1970s by Tim McDonald and Haley Pearsall[121] after their proposal was endorsed "in principle" by the Falls of the James Scenic River Advisory Committee and permission granted by city council to use park property for entering and exiting the river. Their company, James River Experiences, operated until 1986 but did not survive a change of ownership and a period of unusually low water.

Rafting in downtown Richmond. *Courtesy of Riverside Outfitters.*

After a short hiatus and a bid process, the city by ordinance awarded a ten-year exclusive rafting franchise to Richmond Raft Company, established by John Alley and Stuart Bateman. In Richmond Raft's first year of operation, still in the low-water period, the customer count was about nine hundred. In the years that followed, the business picked up in volume, approaching five thousand customers per year, and was featured in a variety of television specials and publications.

In return for the "exclusive rafting franchise" granted by the city allowing the use of the James River Park for launching of commercial rafts, Richmond Raft Company paid the city a percentage of its revenues. This "exclusive" feature of the contract in later years became an issue of controversy. "I think the only thing we had a monopoly on was payment to the city," one of the owners asserted.[122]

Richmond Raft continued its growth curve despite some chafing about the restrictions imposed by the city. At issue was not simply the fees the city charged, but the owners and guides also considered the no customer swimming rule (not strictly enforced), the prohibition of trips at a water level above nine feet and the extra paddle rule all to be unnecessary. Nevertheless, the profile of whitewater rafting within a city grew as did the customer count (although not the profit margin in this labor-intensive business, claimed the owners).

The ownership of Richmond Raft Company changed hands in 1998. Buzz Kraft, through his company Adventure Challenge, had done some booking for Richmond Raft and put a bug in Bateman's ear—"If you ever want out, let me know." Bateman and Alley did, and the sale was made.

The late 1990s, however, was the beginning of a rough patch between the city and commercial rafting as well as the larger paddling community. The initiating event was a Request for Proposals (RFP) asking for bids to provide "Canoeing/Kayaking Franchise along James River"[123] This RFP (note that rafting is not included) began to stir up concern among paddlers who frequented the James. The worry was that it would impose restrictions on private individual paddlers. Although city officials contended that there was no intent to restrict private boaters, emails and calls to the mayor, council members and others let it be known that many paddlers were still concerned. A positive outcome of the friction was the formation of the James River Outdoor Coalition, a group that has gone on to complete multiple volunteer projects and raise funds for the James River Park.

The Canoeing/Kayaking RFP of 1997 was put on hold, but it was followed in 1999 by another for a "Rafting/Canoeing/Kayaking Concession along James River"[124] The occasion for this RFP was the impending expiration of the contract with Richmond Raft Company for the rafting concession and the need to open that contract to a competitive process. A Pre-Proposal Conference for the 1999 RFP was attended by approximately a dozen local outfitters and interested parties. The meeting produced a lively, sometimes heated, discussion about fundamental issues of river policy as much as the details of the RFP.

- Should "commercial" access be limited, thereby justifying an exclusive contract? On what legal foundation?
- What is commercial? Are nonprofit, university and governmental programs that charge a participation fee considered "commercial"?
- Is there an alternative to designating a city concessionaire? Why only one?

As a result of the meeting, the issues raised and the pressure applied, the canoeing and kayaking portion of the RFP was removed and the decision communicated through an addendum to the RFP. The addendum confirmed that only commercial operations would be affected by the concession contract; private boating access, whether in rafts, canoes or kayaks, would not be hindered in any way. The definition of commercial confirmed in the addendum was a basis for city-rafting conflict for years to come. Namely, educational institutions, nonprofit groups and other local governments that charge a fee are considered commercial. "If the operation [of these entities] infringes on the concessionaire, they should not be conducting these operations."[125]

Richmond Raft Company was the successful bidder for the revised RFP and, after a lengthy eighteen-month negotiation, including appeals to the mayor and council to intervene, signed a contract with the city to continue the rafting concession for a five-year period. While the raft company continued to operate with a high volume of customers and trips, all was not well behind the scenes. The contentious relationship between the city and Richmond Raft Company ultimately led to a decision by Richmond Raft to close its business.

A variety of issues plagued the relationship, but the central matter was the contention by Richmond Raft that the city was allowing other commercial rafting operations (those that charged a fee) to use the park without meeting the same obligations and costs the city imposed on it. According to the company, this was a violation of the rafting contract. No satisfactory resolution of the issue took place, and Richmond Raft Company filed a lawsuit in 2005 seeking to enforce compliance. The court heard the case in May 2006 and found in favor of the raft company, but no action followed. Despite the lawsuit and its outcome, Richmond Raft Company and the city entered into negotiations to revise and extend the rafting contract. Multiple obstacles prevented its extension, so Richmond Raft informed the director of the department of parks, recreation and community facilities that it had begun to "close shop."

A period with no commercial rafting followed after Richmond Raft terminated its business, but new entrepreneurs soon began to fill the void. The city moved from a single, exclusive rafting concession to an open process that allows more than one concessionaire (a procedure suggested in the contentious RFP meeting years before). This new "per trip permit system" with a per trip or per person fee to the city allows multiple commercial vendors to provide paddling opportunities. Riverside Outfitters and RVA Paddlesports added raft trips to their other rental and instructional offerings. Several former Richmond Raft Company guides took their experience and skill to these new concessionaires. Current city policy allows up to three raft outfitters to operate. Since the process opened up, outfitters have added many other activities in and along the river: stand-up paddleboarding, tree-climbing, sit-on-top kayaking, trips for persons with disabilities and more.

FISHING

"There are catchable fish in the James River within sight of Richmond's downtown skyline all 365 days of the year." So states John Bryan (who probably has done just that) in his opening sentence in the fishing chapter of his book *The James River in Richmond: Your Guide to Enjoying America's Best Urban Waterway*.[126] And there are a variety of gamefish to go after: smallmouth bass, largemouth bass, sunfish, channel, flathead and blue catfish, as well as migratory fish during spring spawning runs: hickory shad, American shad, white perch and striped bass (aka rockfish). Depending on the target, bait can vary from artificial lures to chicken livers to sunfish, as well as the old standbys—worms and minnows.

The fish one is likely to catch depends in part whether you fish above the beginning of the fall line or in the tidal river below. Of course, the fish do not strictly obey this boundary. Upriver, smallmouth bass and flathead catfish seem most popular; below, blue cats and largemouth bass are more common. Renowned for smallmouth bass fishing, the Richmond area now seems a

Fishing the tidal James in Richmond. *Photo by author.*

Shad fishing in the spring. Some days are even busier. *Photo by author.*

favorite spot for flathead catfish. As the Virginia Department of Game and Inland Fisheries' website proclaimed in 2010, "It's safe to argue that the stretch of James River in Virginia's capital city is tops in the Commonwealth for trophy-sized flatheads—those meeting or exceeding 40 inches in length or pulling down the scales to 25 pounds or more."

Anadromous fish can be found both above and below the fall line, but a prime congregation point is at the uppermost stretch of the tidal river below Mayo's Bridge. To quote Bryan again: "The James River's 400 miles have a single very best fishing location at a single time of year: downtown Richmond in April. That's when and where the hickory shad, American shad, white perch, and stripers congregate on their upstream spawning runs. On April weekends there are many, many boat and shoreline anglers casting lines—so, unless you like elbows, fish on a weekday."[127] Fishing in the tidal river is not just a spring activity however; fishing goes on year-round there as well as at other locations on the river. Regulars are out almost every day. One concern with new downriver plans is the gentrification effect that might push out traditional fishermen and women.

THE RIVER AS MAGNET: HIKING, BIKING, RUNNING, CLIMBING, DOG-WALKING AND MORE

The river, once shunned, is now a magnet for activities of all kinds, many seemingly unrelated to the river. As the photos show, it is a place for catching up on office work (or perhaps writing a novel), painting and even baptism. Of course, that list could be expanded many times over with scenic, wildlife and action photography, guitar playing, poetry writing, wine sipping, riverside partying, salamander stalking and who knows what else.

For many river enthusiasts, it is the trails along the river that are the primary draw. The opportunities to hike, mountain bike, run and walk dogs both in and beyond the park have continued to grow. There are multiple trails and combinations of trails and several loops that cross the river twice and allow return to the starting point without retracing one's steps as well as trails that connect to adjoining neighborhoods. As one trail rider enthusiastically proclaimed, "How many trail systems offer what JRPS does? You've got a trail with a wicked personality disorder—technical rocky/rooty in some sections, flowy in others, creek crossings and

A river is more inspirational than an office. *Photo by author.*

Riverside as art studio. *Photo by author.*

An authentic place for baptism. *Photo by author.*

Perhaps this is part of the trail that has "wicked personality disorder." *Photo by author.*

tree dodging, awesome views of urban skyscrapers cut by a river offering class IV and V rapids, mixed with kudzu jungles and kamikaze chipmunks and groundhogs. The close proximity to an awesome coffee house and brewery don't hurt either!"[128]

Other users are not so interested in the "wicked personality disorder" and prefer the more serene. There are plenty of trails, some within the park and some not, that meet this need, including the Canal Walk, the Virginia Capital Trail, Slave Trail and the T. Tyler Potterfield Memorial Bridge. Almost every section and loop of the riverside trails has regular dog walkers or others just out for a stroll, alone, in pairs or small groups. Dogs seem to be a favorite excuse for their owners to take to the highly popular riverside trails, and most do pick up after their pet's nature call.

Walking clubs frequent the riverside trails, including one with James River in its name, James River Hikers. The trails also are home to a variety of competitions, both mountain biking and running. Those who are looking for a vertical challenge regularly head to the Manchester Climbing Wall on the abutment of the long-abandoned Richmond-Petersburg Railroad Bridge.

James River Hikers about to cross T. Tyler Potterfield Memorial Bridge. *Photo by author.*

It is not just trails in the park that attract hikers, bikers, runners, dog walkers and stroller pushers to the river. Riverside Drive on both sides of Huguenot Flatwater has more recreational activity than vehicular traffic, and some of the vehicular traffic is recreational—folks out for a drive to take a look at the river.

Youths are not left out of river activities. The river, with Belle Isle as the center of action, is the attraction for Passages, a popular youth day camp for both boys and girls that features kayaking, rock climbing and mountain biking as well as zip-lining across the quarry pond. The quarry climbing wall, the pond, the mountain bike skills area and First Break Rapid are beehives of spirited play when camp is in session. The first lesson taught and reinforced throughout each weeklong session (and one appreciated by parents) is "No Whining."

Take a look at Belle Isle by Hollywood Rapid or Pony Pasture on a summer weekend, and it becomes evident that rocks are as suitable for sunbathing, picnicking and lounging as a sandy beach (and there are even sandy beaches along the river). And why not hang a hammock and doze to the music of the rapids? Some visitors even get in the water. Snorkeling is

Left: Passages campers at work on the Belle Isle quarry wall. *Photo by author.*

Below: Dozing to the music of the rapids. *Photo by author.*

Dragon boat races have become an annual competition in the tidal James near Rocketts Landing. *Photo by author.*

popular with some, especially when the water is low and clear, and a few hardy souls consider long-distance swimming upstream against the current an enjoyable activity. Leisurely tour boat rides with historical commentary are available in season on the restored canal, as are eagle and sturgeon viewing tours downriver.

The river setting, especially the Brown's Island area, is a popular spot for a variety of festivals and gatherings offered by Venture Richmond and others. Friday Night Cheers, music and beer, is a staple in late spring and early summer. Dominion Riverrock is a music and sports festival that includes a mud run, mountain biking and kayaking. The Dragon Boat Races sponsored by Richmond Sports Backers and James River Advisory Council's Parade of Lights have become annual events in the downriver reach. The Richmond Folk Festival, too, is held on the riverfront in the new amphitheater carved out of the hillside (amid some controversy) at the edge of the Kanawha Canal bed near Tredegar.

PERSPECTIVE

There is no doubt that play along the river has mushroomed over the years. At one time, only those who owned or were willing to trespass on private property could enjoy recreation along the river. The number and types of players on and along the river has continued to grow year after year, and the Riverfront Plan implementation and James River Park growth likely will increase the numbers by opening new opportunities. Park and river use rules have been sorted out. The high-water 5/9 rule is well accepted, even though questions about the legality of its enforcement have been raised, and multiple outfitters now offer opportunities from rafting to kayak instruction, fishing and eagle and sturgeon viewing tours.

Attitudes, opportunities and activities regarding play on the river have changed considerably over the years as the river has been transformed from abuse and neglect. City officials suffered some growing pains during the process as they became more accustomed to river activities that were outside their training and experience. That change is dramatically illustrated with the shift from a park superintendent prohibiting a boater from launching his kayak in 1975 to a councilwoman in 2015 inviting the public to enjoy a whole litany of river activities—and her list could have been even longer.

Of course, some liked the river better when fewer persons found their way there. Some preferred to make their own trails or to paddle when no one else was around. There are still times and places one can find seclusion, but these times and places are less plentiful than they once were. Most users seem to think the expanded opportunities for access and types of play are worth the tradeoff. No doubt issues and conflicts will emerge as time goes on, but the pattern of active use of the river as an amenity seems firmly set.

———⟡———

The establishment of the James River Park, largely on land that had seen no development, provided an inviting venue for the rapid growth of recreation on and along the river. Some visionaries had big ideas about the recreational and commercial potential for an area that had been developed and redeveloped many times—the downtown riverfront.

7

RIVERFRONT

A VISION THAT LINKS THE FUTURE TO THE PAST

Today the James is a "diamond in the rough" begging to be polished as Richmond's showpiece.
—Richmond and the James, *City of Richmond Planning Commission, 1967*

At some point—who knows when?—Richmond remembered the value of the jewel in its midst....Great accomplishments depend on a vision. And the vision of a Richmond centered on the James pleases the eye and invigorates the spirit.
—Richmond Times-Dispatch *editorial, July 26, 1996*

Despite pollution and the historical clutter along its banks, the potential for making the downtown riverfront something special had been recognized for decades. Beginning shortly after mid-century, various organizations and individuals began to build a vision of what the downtown riverfront could be. Some of these efforts were officially requested plans; others were unofficial discussions and writing, some preserved and some not. Several newspapers, including the *Richmond News Leader*, the *Richmond Times-Dispatch*, *Style Weekly*, *Metropolitan Observer* and the short-lived *City Edition* participated in these periodic discussions. Many early visions of what the river might be were not implemented immediately, or not at all, but they did begin to shape thinking and attitudes about the river.

The riverfront in the central business district is quite different from the less disturbed reach between Bosher's Dam and Lee Bridge. The shoreline downtown was highly developed before the modern transformation of the river began after mid-century. Although this shoreline is no longer a candidate for wilderness, it still has pristine islands and some of the best whitewater paddling and fishing anywhere.

The downtown banks of the river have been developed and redeveloped many times. It was here that ships coming upstream were stopped by the "falls" and here that waterpower was used to mill flour, make rope, produce iron products and generate electricity. Over time, these activities and structures have been replaced by financial institutions, service industries, restaurants and high-rise residential structures.

During the modern transformation of the river, orienting, or reorienting, the city toward the river has been an overarching and recurring theme. During the time when transportation and water power were vital to the economy, the river was a major focus of attention and economic activity in the city. Afterward, the city largely turned away from the river, considering it an obstacle as much as a resource. As Brenton S. Halsey, a major figure in transforming the downtown riverfront, remarked: "In the 20th Century, Richmond turned its back on the river and buried its riverfront in dilapidated industry, railroads, highways, and general neglect—exacerbated by frequent flooding."[129] The central theme of plans

Brown's Island as it stood in 1978, now an open meadow for festivals, concerts and the display of history. *Courtesy of Cabell Library Archives, Virginia Commonwealth University.*

Part of the riverfront as cleanup and renovation began in the 1980s. *Courtesy of Venture Richmond.*

during the last fifty-plus years, both implicitly and explicitly, has been to undo the period of neglect and again make the river the centerpiece of the city—a centerpiece with a new purpose.

AN EARLY RIVERFRONT PLAN: THE RIVER AS CITY CENTERPIECE

A plan prepared in the 1960s by the Richmond Planning Commission captures what might be considered a starting point for post-industrial riverfront transformation in downtown Richmond. The preamble of *Richmond and the James: A Plan for Conservation, Recreation, Beautification* captures the status of the river at the time of writing as well as aspirations for it: "Human wastes and industrial pollution are vexing problems, prohibiting development of the recreational potential the river has to offer. Nevertheless, her wilderness qualities remain and her 'white waters' are found refreshing by the intrepid angler who trespasses to reach her banks....Today the James is a 'diamond in the rough' begging to be polished as Richmond's showpiece."[130]

The plan featured interconnecting trails while augmenting the beauty of the natural setting "where necessary to correct mistakes of nature as well as those of man." The central theme, the plan stated, will be "studied naturalness" and "reorientation of the City's development so as to center on the river." The plan's elements featured use of City Dock as a public marina, restoration of the canal basin and locks, use of the canal for leisurely barge rides, emphasis on Belle Isle's historical significance and preservation of the area as a wildlife sanctuary. (A plan that overlapped and reinforced the Planning Commission concept was Abbott's 1968 master plan for the James River Park which also foresaw Belle Isle as a prominent element; see chapter 3.) The commission's plan also called for low dams across the river and between islands that would serve as footbridges and at the same time raise the water level to create "suitable boating and fishing areas."

Except for some specific items—the dams, for example—this plan set a theme for many plans and projects that followed. Especially noteworthy is the effort to make the river the focus of the city, something that clearly was not the case at mid-century. The river was to be the focus, the centerpiece, of Richmond, but the canal was to be the link to the river. Soon after this vision of the river as centerpiece was presented, a severe threat arose to that key link.

THE CANAL AS LINK

The James River and Kanawha Canal, often just called the Kanawha Canal, was the most ambitious capital project undertaken in eighteenth- and nineteenth-century Virginia. Financing and building the canal, despite the support of such luminaries as George Washington, John Marshall and Joseph Cabell, was a constant struggle. Money problems, regional jealousies, floods, the Civil War and competition from railroads were all challenges to the canal's construction, maintenance and profitable operation. It reached its heyday in the 1850s only to then become a victim of nature, war and the railroad. For almost one hundred years prior, great effort had been made to build and operate a canal along the James River and beyond. Ultimately, the canal reached Buchanan, 197.5 miles west of Richmond, but the dream of a water route to the Ohio and then Mississippi Rivers was never achieved. The canal, as a means of transport, was abandoned in 1880, when it was sold to the Richmond and Allegheny Railway; rail, rather than water, had become the preferred mode of transport.

Nearly one hundred years after the rail company had acquired the canal and discontinued its use as water transport, historians and canal enthusiasts began an effort to preserve for posterity what remained of the Kanawha and other canals. The vision of canal enthusiasts in the 1960s and early 1970s was to return the entire length of the canal in Richmond to working order for recreation and educational purposes. Such an effort nicely matched the aspiration for a renewed and accessible riverfront. Just as building the canal had been a struggle, however, so was the task of preserving it—a task not without complication. As is usually the case with historic preservation, there were new competing uses. Three threats to the preservation and restoration of the canal in the downtown area arose in the 1960s and '70s: a downtown expressway, high-rise buildings and the proposed use of the Richmond Dock (the final canal link to the river) as a sewage retention basin (see chapter 2).

At that time, the length of the canal in Richmond was intact or capable of being restored. One of the early voices for preservation was that of Dr. William E. Trout III, a Richmond native who at the time lived and worked in California. He began a several-decade avocation documenting and advocating for the preservation of canals. In the 1966 paper "The James River and Kanawha Canal: Its Potential as Part of Richmond's River Recreation Area," Trout laid out the opportunities for preserving, restoring and using the canal in a new way.[131] In a document a few years later, Trout confirmed that preserving the canal was still viable: "All 10½ miles of the Canal are almost completely intact. There are nine locks in excellent condition, suitable for restoration to working order. These are the Ship Lock, the five Tidewater Connection locks, the two 3-mile Locks at Byrd Park, and the 5-mile Lock at Dead Man's Hill."[132]

So, at the time that reviving the riverfront began to attract some interest and gain momentum, the Kanawha Canal was still viable for preservation. Much of the canal, from Bosher's Dam to a point near Tredegar, was watered since it had been taken over by the city as an intake canal for its water supply and treatment system. In downtown Richmond, from Tredegar to about 12th Street, the canal had been filled and, in some places, covered with asphalt. The Great Turning Basin had been converted to a railyard and later a parking lot, but the stonework was thought to be relatively intact underneath. Three of the five locks between the Basin and Richmond Dock, called the Tidewater Connection, were covered but still capable of restoration. These five locks had lowered and raised boats sixty-nine vertical feet in a distance of three and a half blocks. Two of these locks, along with the Richmond Dock and the Great Ship Lock, were in working order.

One of the five Tidewater Connection Locks in 1968. *Courtesy of the Library of Congress.*

Trout was not the first nor was he alone in expressing hope for canal preservation and restoration. Dr. Bruce English, president of the Historic Richmond Foundation, said in 1967, "It wouldn't take much to make a considerable portion [of the canal] work again." An officer of the Canal Society of New York wrote: "Surely the people of Richmond—who in this matter must act as trustees for all the people of our country—will not permit one of the most historical sites on the Eastern Seaboard to be destroyed."[133] Others joined the call to preserve the canal, including a number of prominent women in the community. Mrs. Eugene Sydnor, Mrs. Cabell Tabb and Mrs. Douglas Rucker held meetings with business leaders and garnered publicity for the cause.[134]

The James River and Kanawha Canal Parks Inc., a group headed by Eugene B. Sydnor Jr. with an impressive list of businessmen, citizens and river enthusiasts as officers and advisory board members, was formed to protect the canal from the rising threats it faced, especially the "Downtown Tollway as Proposed by the Richmond Metropolitan Authority." The organizational purposes listed in its charter were:

The Great Turning Basin as a parking lot, 1968. *Courtesy of the Library of Congress.*

- The protection of the James River and Kanawha Canal in its original route.
- The restoration of the canal and any related structures.
- The creation of adjacent walkways and ribbon parkways.
- The establishment of such a canal parks system as an integral part of the total development of the James River Park System.[135]

Unlike the fight against the Riverside Parkway on the south side of the river (chapter 3), the attempt to alter the route of the Downtown Expressway away from the historic canal was not successful. Many people thought the expressway would serve its purpose of bringing people downtown if it terminated at 7th Street, thus saving much of the historic canal, including several locks. The Wilbur Smith traffic study, however, concluded that a connection to I-95 was necessary to generate enough traffic to justify the expressway. The members of the newly formed Main to the James Development Committee of landowners and city officials formed by Mayor Thomas J. Bliley Jr. in 1971 were initially divided but, following the lead of city manager William Leidinger and the traffic consultant, voted in favor of the full Downtown Expressway.[136]

By the spring of 1974, three of the five locks of the Tidewater Connection were being dismantled.[137] As a nod to historic preservation, the stones of the locks were numbered to allow historically accurate reconstruction and stored at another location. Although the stones were labeled, it was done with paint that washed off, and some of the stones still remain stored under Manchester Bridge. Nevertheless, many were in fact used in the Canal Walk project years later.[138] Bottom line, however, the potential for reopening the original route of the canal in downtown Richmond was gone.

The two easternmost locks on the property of Reynolds Metals were preserved, however. The additional warehouse space required by the Reynolds Packaging Division was designed to preserve the locks by building over and around the locks on the property. In this way, the two locks were made accessible to the public and preserved for the city to restore the locks to working order in the future.[139]

The high-rise threat also came to pass. The Great Basin was not restored, and a decade later the parking lot that had covered the basin was replaced by

One of the preserved Tidewater Connection Locks, 2019. *Photo by author.*

the James Center—three towers of mixed-use offices and commercial space. The remains of several dozen canalboats were unearthed in the construction process under the supervision of archaeologists, but all that now remains of the Great Turning Basin are plaques commemorating it. A hotel and two office towers now stand on the site.

The third threat was averted—the Richmond Dock did escape use as a retention basin for sewage overflow. The objections were strong, and an alternative, the Shockoe Retention Basin, met the need (see chapter 2).

The loss of the downtown route of the Kanawha Canal was a disappointment to canal enthusiasts and a setback for the riverfront planning that had gone before, but the hope for a new riverfront had not been lost, nor had the hope for a restored canal, albeit one that deviated in places from the canal's historic route. There was barely a pause before discussion, planning and hopes for a revitalized riverfront resumed.

THE RIVERFRONT VISION CONTINUES

In 1977, a few years after the three canal locks were dismantled, reorienting the city to the river again was promoted in a new plan.[140] *Vanishing and Returning Gardens* echoed the planning commission document from a decade earlier. This ambitious plan proposed that the James in Richmond become a national attraction by aggressively showcasing the river. The "vanishing and returning gardens" theme was based on seasonal fluctuations in river levels because high water in spring hides rocks and islands and low water in summer returns them to view. The intent of the plan was to "reorient the city to the river," with the islands between the Manchester Bridge and Mayo's Island serving as the focus for the plan. After removing fallen bridge supports, a "river promenade and some of the islands will be connected by paths and stepping stones." These connections would be designed "not to disrupt the natural flow of the river, and to disappear during floods." The key feature was "James Gate" composed of terraces and a parking structure at the north end of Manchester Bridge from the Federal Reserve to the riverfront. The terraces would have seating, panoramic views up and down the river and restaurants on various levels. This would be the link between downtown and the river.

Other features of this plan are also noteworthy. Tredegar would be a historic exhibit, and an amphitheater would be carved in the hill behind it. The floodwall, then in serious discussion, would be designed as a sculptural element, with landscaped terraces, promenades and overlooks. Belle Isle would be left natural except for the area east of Lee Bridge. The eastern tip of the island would contain a ferry landing, beach and picnic tables. Mayo's Island would be devoted to public use with a picnic area and a children's park; private enterprise would be located elsewhere. The forthcoming retention basin on Chapel Island would hold tennis and other court games on its roof, and the adjoining City Dock would have a new marina with access to the river through a functioning Great Ship Lock. Canals would be reconstructed where possible with boat rides to the Pump House and back. And to bring it all together, all points within the "River Park" would be connected by walkways, promenades or open spaces.

The authors of *Vanishing and Returning Gardens* acknowledged that the plan was ambitious, indicating that it was "virtually a new town in-town" and would take fifteen to twenty years to implement. It proved more ambitious than the city was willing to tackle.

DESIGNING A RIVERFRONT AROUND A FLOODWALL

The floods from Hurricane Camille in 1969 and Tropical Storm Agnes in 1972 had stimulated serious pursuit of funding for a floodwall (see chapter 8), so the design of a floodwall in the downtown riverfront entered the conversation. A study by Glave Newman Anderson & Associates in 1979 sought to integrate the proposed, but still unfunded, floodwall with a continuous riverside park featuring improved access to the river from Belle Isle to the foot of Church Hill.[141] The central concept, reminiscent of plans that had gone before, was a continuous riverfront park that stretched from Belle Isle in the west to the foot of Church Hill in the east. This proposed park would include Belle Isle, Mayo's Island and Chapel Island, as well as much of Brown's Island. Improved access to the river was prominent in this plan, including paths that line both sides of the river. Parallel paths were recommended on the north side of the river, both to run from Tredegar to Great Ship Lock. One path would run on the north side of City Dock and Brown's Island; the other on the south side of Chapel Island. A third walk would circumnavigate Mayo's Island and include extensions to some of the islands west of Mayo's. At the northern end of Mayo's Bridge, a major esplanade would be created in the river itself. This would be at the base of the twenty-four-foot-high nineteenth-century granite retaining wall, with the lower level connected to the upper by several stairs. The location and character of the floodwall would be designed to blend with the park-like riverfront that was envisioned.

The Richmond First Club added to riverfront discussion in April 1984 with a report intended to stimulate discussion and serve as a springboard for action. *The Richmond First Club Report on Metropolitan Richmond's James River Corridor* gave primary attention to the riverfront from Oregon Hill to Tobacco Row but was not restricted to this area. Richmond First revived several recommendations that had been made before and added some new ones. Once again, design control for the upcoming floodwall was emphasized, as was restoration of the canal from Byrd Park Pump House to Ship Lock.

CANAL VISIONARIES BACK AT WORK

Prompted by the threat posed by the forthcoming construction of the floodwall, a citizen Canal Committee was organized and chaired by A. Howe Todd, former planning director and assistant city manager. Placed administratively under the Historic Richmond Foundation, a central task of the committee was to find a way to preserve the surviving structures of the original canal system. Through a grant from the Richmond Industrial Development Authority, the committee hired Carlton Abbott and Partners to study the options available.[142] The 1988 plan that Abbott and the committee produced was a significant refinement of the riverfront vision. Halsey calls it "the most elegant, creative and comprehensive plan for the restoration of the canals that had heretofore been created."[143]

Since part of the original route of the Kanawha Canal was no longer available for restoration, the Historic Richmond Foundation Canal Committee recommended an idea that had been suggested during the earlier downtown expressway controversy—using the parallel Haxall Canal as a connector between the remaining Kanawha Canal location near Ethyl Corporation headquarters to the west and Richmond Dock to the east. This historic route ran through an undeveloped portion of Ethyl property, so to make the recommended connection to the remaining Kanawha Canal bed required the concurrence of the Ethyl Corporation. Ethyl had earlier filled in that portion of the canal, but it could have been restored without difficulty. Ethyl was the focus of considerable lobbying, negotiation and even pleading. An example of an effort at persuasion is

The bed of the James River and Kanawha Canal near Ethyl Corporation in 1990. *Courtesy of Cabell Library Archives, Virginia Commonwealth University.*

found in a letter from R.B. Young to Floyd Gottwald, Ethyl president. "I feel strongly that you now have a unique opportunity to create an even stronger enhancement of your corporate image," Young argued. "Ethyl Corporation could become one of the most highly praised corporate citizens of our state should you choose to incorporate the Kanawha Canal as the centerpiece to your development in this area."[144]

Despite the efforts at persuasion, Ethyl refused to allow the canal reconnection the committee preferred, although an alternate connection route that would involve a smaller and less prominent part of Ethyl's property was accepted. So, the route that would have allowed the most direct connection from a restored Haxall Canal to the bed of the Kanawha Canal, and one that would have preserved more of the historic canal as well as an iconic view of the city, was not to be.

Canal advocates, of course, were not pleased. Lyle E. Browning, president-elect of the Archeological Society of Virginia, is quoted as saying that a highway marker should be erected that states: "Near here ran the James River and Kanawha Canal, designed by George Washington and removed by the city of Richmond to allow a developer to build an office complex."[145]

Support for the proposed alternate route was lukewarm at best. Several organizations opposed the compromise route, as did the city planning commission, at least initially. The route proposed by the canal committee was part of the city's downtown master plan, so the compromise was placed on the planning commission agenda. The commission first rejected the compromise amendment to the plan. City council then passed the change over the objection of the planning commission, but the commission later approved it at a meeting with more members present.[146] Although unpopular with many, the alternate route did preserve the possibility of a Haxall-Kanawha connection and for a navigable canal running west to Maymont and beyond. As of this writing, that alternate route has not been completed, but continued canal restoration is not dead. Venture Richmond is pursuing the possibility of restoring the Kanawha Canal from Maymont to Tredegar Green, just west of the former Ethyl property.

Canal enthusiasts were successful a little farther downstream where the canal in the form of Richmond Dock meets the James. With another grant from the Richmond Industrial Development Authority, the Great Ship Lock was restored to working order. To symbolize the achievement and to publicize canal restoration potential, "the tall ship *Alexandria*" sailed up the river and was lifted to canal/dock level in November 1989.[147] The Norfolk Southern bascule bridge near the lock, which blocks continuation into the dock, was

The James River and Kanawha Canal just east of the Pump House, 2019. *Photo by author.*

Above: The Great Ship Lock, the final connection to the tidal river and route to the Chesapeake Bay and Atlantic Ocean, 2019. *Photo by author.*

Left: The schooner *Alexandria* in the Great Ship Lock during a ceremony by Historic Richmond Foundation, 1989. The lock was put in working order, if only temporarily. *Courtesy of Historic Richmond Foundation.*

not opened. Although the lock has not been maintained in working order nor an arrangement made with Norfolk Southern regarding the bridge, lifting the ship into Richmond Dock dramatically symbolized the vision of a restored canal that had been building for several decades.

PERSPECTIVE

Even while the riverbanks were cluttered with remnants of Richmond's industrial past and the river only beginning to lose its function as a "sewage treatment plant," visionaries saw the potential for the James River, not just the park upstream but also the developed area downtown. By the late 1980s and early '90s, those plans had been massaged multiple times and interest in the downtown riverfront had reached a take-off point.

Viewed now, these several plans of the 1960s, '70s and '80s invoke many reactions: that concept was implemented, has served us well and made future steps more feasible (use of Haxall Canal as a connector); that concept may have sounded good at the time, but we sure are glad that it was never carried out (ice skating rink on Belle Isle); they tried that idea, but floods have shown it to be impractical if not impossible (series of footbridges linking islands); that idea was a good one and perhaps now we have the political will to carry it out (making Mayo's Island a signature downtown park); and more. Virtually every plan proposed to make the James River the centerpiece of the new Richmond.

While the river was to be the centerpiece for the city, the restored canals continued to be thought of as the connection to the river or in some cases as an objective in its own right. Every plan envisioned some role for a restored canal. By the end of the 1980s, hopes for a revitalized, modernized and accessible downtown riverfront had been growing for three decades. Some on-the-ground action had occurred by then, but more importantly the aspiration had been established.

While many advocates worked with energy and passion to preserve the canals and open the riverfront to the public, others labored just as energetically and passionately to prevent the next big flood from damaging parts of the highly developed downtown.

RIVERFRONT

FLOODS AND A WALL

Camille had been the ultimate storm. Nothing could approach the havoc dealt
by that treacherous lady. There were many who refused to believe that the James
could possibly rise beyond the 1969 level. Until Agnes.
—*James Berry,* The Richmond Flood

Note that the intent of landscaping the wall is to try to soften it and to help [it] *to*
disappear where possible.
—*Carlton Abbott and Partners,* Richmond Floodwall Enhancement, *1991*

During the same period that the vision for opening the downtown riverfront matured, a parallel plan for that riverfront moved forward. Historically, floods had been a defining feature of the James River; in 1994, a floodwall became a defining element of the riverfront in much of downtown Richmond. Although the James River in Richmond had flooded multiple times during its recorded history, it is fair to say that the 1972 flood precipitated by Tropical Storm Agnes, coming as it did three years after the Hurricane Camille flood, was the catalyst that brought about flood control action.

THE FLOODS

The flood of May 1771 sometimes is and sometimes is not listed as the highest to hit Richmond. The markers on the floodwall put it at the top, but other listings estimate that it reached a height less than Agnes. Nevertheless, it was a major, damaging flood. As Dabney describes the account of one witness in *Richmond: The Story of a City*: "When the deluge reached full volume, dwellings came scudding on the surface of the raging river, along with warehouses, wine casks, hogsheads, trees, lumber and cattle. From the roofs of some of the fast-moving houses terrified persons called for help."[148] The exact height, of course, is impossible to know. Floods occurred periodically after 1771, but for two hundred years none challenged it for volume and velocity.

HURRICANE CAMILLE, 1969

Hurricane Camille began its entry into the U.S. mainland along the Mississippi coast with 200-mile-per-hour winds, the strongest to make landfall in the United States in the twentieth century. It moved up the Mississippi Valley for 150 miles, turned east and entered Virginia as a tropical depression. Settling in the Blue Ridge south of Charlottesville, Camille spawned, as the Corps of Engineers called it, "one of nature's rare events."[149] As Swift describes it in *Journey on the James*, "[Rain] fell so hard that the ground turned to pudding and swallowed up forests and farms. Birds drowned in the trees. In the flash

Viewing the impact of the Hurricane Camille flood, 1969. *Courtesy of Richmond Times-Dispatch Collection, The Valentine.*

floods and avalanches that followed, 125 people were drowned, crushed, or buried so deep they'd never be found."[150]

Richmond did not have it quite so bad. The crest of muddy water moved down the river, allowing Richmond to prepare as well as it could. The city had been warned, but little could be done to prevent the property damage that followed. Fortunately, there was no loss of life. The crest arrived downtown on August 22 and reached 28.6 feet at city locks.

TROPICAL STORM AGNES, 1972

Agnes was only a weak hurricane when it formed in the Gulf of Mexico and had weakened even further to a tropical storm when it arrived in the Mid-Atlantic, seemingly a rather innocent weather event. As it moved across Virginia, however, Agnes collided with another storm, moved inland, slowed and produced the most damaging flood in Richmond's history. It wreaked widespread havoc along the Mid-Atlantic, especially in Virginia, Maryland and Pennsylvania; the James crested in Richmond on June 22, almost eight feet higher than Camille.

> *Friday, June 23, disaster turned into calamity. At 7 a.m. the city ordered the downtown area sealed off. Floodwaters had covered the approaches of key bridges serving the city and a major power failure occurred. Barricades went up, manned by the police and National Guardsmen. The capital of Virginia took on the look of an occupied city. Only armed troops, some of the 1,800 called out, were in sight. Downtown Richmond stood almost deserted on a bleak drizzly Friday that got progressively bleaker.*[151]

The volume of the Agnes flood exceeded all that had gone before (with the possible exception of the flood of 1771) and all that have occurred since. Camille, it had been thought, was the ultimate. "Camille was what the experts called a 100-year-rain. Nothing like it in recorded history. Nothing like it probably ever again. At least not for 100 years."[152] Surprise. Three years later, Agnes arrived. "The heavy flooding was to produce the highest cresting of the James River in Richmond history—36.5 feet at 4 p.m. Friday, June 23. In 1969 Camille had reached 28.0 feet and at the time there were those who believed firmly that the James could never possibly go any higher."[153]

Agnes floodwater, 1972. *Courtesy of Tricia Pearsall from the Dale Wiley Collection.*

Estimated damage from Camille was $23 million and from Agnes $59 million, for a total of $82 million in less than three years.[154] To add an exclamation point to the catalyst for action, an additional flood in 1972 occurred in the fall. The *Richmond News Leader* headline put that flood in perspective: "It Isn't Camille or Agnes, But…"[155] And more perspective added by the *Richmond Times-Dispatch* the following day: "First There Was Camille, Then Agnes. Call This One Routine."[156] That flood crested at something over twenty-four feet.

Little time was lost after the Agnes flood before action began on two fronts. The U.S. Army Corps of Engineers began a feasibility study to determine how flood protection might be achieved, and the Central Richmond Association Flood Protection Committee was formed to advocate for a floodwall. The immediate aftermath, however, was the cleanup. Thousands of hours were spent shoveling mud from storefronts, removing mildew and all the other tedious chores involved in cleaning up after a flood.

THE CORPS STUDY

Not long after the Agnes flood, the U.S. Army Corps of Engineers launched a feasibility study to determine ways to mitigate future flood damage. As the corps report stated, "Flooding experienced by the city of Richmond, due to Hurricane Camille, Tropical Storm Agnes, and the October 1972 flood has brought concern over the serious flood problem....This concern prompted the Corps of Engineers to expedite a study of flood protection for Richmond."[157] The corps had a study of the entire James River Basin already underway that had been authorized by Congress in 1964.

The 1974 Army Corps of Engineers study, completed two years after the Agnes flood, considered a total of fourteen options for flood control in Richmond, eight nonstructural and six structural. The nonstructural options were:

- Evacuation of the floodplains
- Flood proofing of buildings
- Combination of flood proofing and evacuation
- Voluntary property sales plus warning clause in future deeds
- Flood insurance
- Floodplain zoning regulating the use and development of floodplains
- Flood forecasting with temporary evacuation
- No action[158]

Each of these nonstructural solutions was considered inadequate or had an unfavorable benefit-cost ratio. Some of these, flood proofing of buildings, flood insurance, floodplain zoning and improved flood forecasting, were considered "good supplemental measures which could be implemented by local interests."[159]

The six structural alternatives considered were:

- Large dams (economically and environmentally unfeasible)
- Small dams on multiple tributaries (not economically feasible)
- Catch basins (not effective)
- River channelization (exorbitant costs and significant environmental damage)
- Channel diversion ("unreasonable and not worthy of further consideration")
- Various floodwall and levee plans[160]

The last of these options, a floodwall, was thought to be the best option. It "would meet the needs of Richmond and would have only minor social, economic, and environmental impacts on the area."[161] The corps report and its recommendation was a critical first step in a long process.

ARGUMENTS FOR AND AGAINST A FLOODWALL— MOSTLY FOR

After the Agnes flood, led by the Central Richmond Association Flood Protection Committee, arguments in favor of constructing a floodwall in Richmond were loud, frequently heard and straightforward. City council quickly lent its support with a resolution adopted on June 24, 1974, and the corps built a strongly worded argument for the proposed action in its environmental impact statement: "If flooding is not controlled, downtown Richmond will continue to be subjected to floods at irregular intervals with the possible loss of human lives and the untold misery and suffering which cannot be measured in monetary terms."[162]

The factual foundation of the pro-floodwall argument was, of course, the number and severity of recent floods. As Dale Wiley, chair of the floodwall committee and "Mr. Floodwall," as he became known for his strong advocacy, put it in one publication:

> *The August 1969 flood, known as Camille, was the worst in modern history and statistically would happen only once every 100 years. Less than three years later, in June of 1972, the Agnes flood occurred and it was again the worst in modern history....Between 1969 and 1987, eight major floods occurred in Richmond and included the first, second, third, and seventh worst floods in modern history.*[163]

The next step in the argument, of course, was the risk of similar damage in future floods. One estimate was that the eight floods since 1969 had caused $185 million in damage in 1987 dollars. With no flood protection, property values would decline, new investments would not take place and the city's tax base would decline.[164]

Congressman Thomas J. Bliley Jr. echoed that argument: "The Richmond floodwall has proven very popular in Congress. It is an investment that will pay off in nearly every way: a historic area will be redeveloped; jobs and investment will be created; and taxes from new commerce will flow into federal, state, and local coffers. In addition the Richmond floodwall has no environmental liabilities."[165]

Even the river seemed complicit in support of a floodwall. Speaking of the 1985 flood that occurred largely as a result of Hurricane Juan, one commentator noted: "If you favor a floodwall, then this flood was perfect timing." The *Richmond News Leader* added, "As the mighty James overflowed its banks and caused heavy damage in the city for the third time in 16 years, the U.S. House of Representatives was ready to vote on a major water resources bill that would provide $94.5 million to build a floodwall in Richmond."[166]

Arguments against constructing the floodwall were rare and rather muted in comparison to the continuing advocacy in favor. Newton Ancarrow, one of a few, took a hardline approach and argued vigorously against the floodwall project. His position was that owners in the floodplain should be given a choice of leaving or staying and taking their chances with nature.[167] Early on, the U.S. Bureau of Sports Fisheries and Wildlife indirectly argued against the floodwall by enthusiastically endorsing "evacuation of the floodplain" as the preferred alternative.[168]

The Conservation Council of Virginia seemingly took a position opposing the floodwall but, in virtually the same breath, backed away from that opposition.

> *The Council must take the position that in the long run evacuation of buildings and businesses in the flood plain is not only the most environmentally sound but also the most effective means of protection…. Evacuation should be carried out at least to the extent that it is feasible, possibly through voluntary sales to the City or State, with the land returned to open space.*

Two paragraphs later in the same position statement, after noting the possibility of other alternatives such as dams and catch basins, the council states its compromise:

> *The Council will support, in theory, the proposed walls and dikes in Richmond if and only if the Corps of Engineers, the Richmond Chamber of Commerce, and the Central Richmond Association will pledge that if the proposed walls and dikes are built, they will not in later years request impoundments in the upper regions of the James River and other extensive public works outside the City as a further cure for Richmond.*[169]

A few city officials raised mild objections. Charles Peters, planning director, argued that it would be a mixed blessing—that it would encourage development but could potentially create a "permanent eyesore."[170] In 1986, at the same time that the Richmond City Council approved its portion of the funding for the floodwall, A. Howe Todd, acting city manager, expressed reservations on the basis of cost, but not opposition to the plan: "I cannot fully recommend this agreement because I do not know the ultimate costs to the city. At the same time, I cannot oppose it because a large segment of the commercial and industrial community needs it so badly."[171] His objection was not to the wall but to the possible cost.

At one point, the Corps of Engineers almost abandoned the project when its benefit-cost analysis showed that costs outweighed benefits. The city, however, successfully argued that the corps was using old data, and a new calculation with fresh data showed a ratio greater than one, putting it in the acceptable range.[172]

MOVING FORWARD

Political support for a floodwall, both locally and nationally, far outweighed the opposition. So, for politics, matters moved ahead at a relatively fast pace.

- In 1976, Water Resources Development Act authorized the Corps of Engineers to proceed with studies.
- In 1980, the Phase I Advanced Engineering & Design Report was completed.
- Congressional authorization for construction was given in October 1985 through the Supplemental Appropriations Act.
- In June 1986, the city and the corps signed an agreement to build the floodwall.
- In November 1988, the first floodwall contract was awarded.[173]

Construction of the floodwall proceeded with only a few hiccups—among them discovery of contaminated soil that delayed completion and added cost, need for congressional appropriation of additional funds and the need for a hole in the wall. Nevertheless, with the construction of the floodwall underway, attention again turned to its appearance—and to that needed hole.

The floodwall under construction. *Courtesy of Tricia Pearsall from the Dale Wiley Collection.*

AVOIDING A "PERMANENT EYESORE" AND CUTTING A HOLE IN THE WALL

Well before the floodwall had received approval and funding, the visual and river access impact had been considered in a study by Glave Newman Anderson & Associates with the guidance of a steering committee that included city manager Manuel Deese and Dale Wiley representing the Central Richmond Association. The *Riverfront Flood Protection and Development Study*, published in 1979, included design recommendations intended to tastefully incorporate the floodwall into a publicly accessible riverfront (see chapter 7). The U.S. Army Corps design did not follow the guidelines and recommendations of the *Riverfront Study*. Opportunities were lost or, at best, postponed.

As the floodwall was being built, a Floodwall Enhancement Committee was formed with Tricia Pearsall as chair. When the enhancement project began in January 1990, the design of the floodwall was almost entirely fixed so the committee was limited to cosmetic treatment. Carlton Abbott and Partners was contracted for design assistance. The purpose of the committee was to find ways to soften the visual impact of the floodwall without affecting the structural integrity of the wall.[174] "The thing we don't want is for the wall to become an eyesore," said Bette Dillehay, Enhancement Committee member.[175] Suggestions included use of trees, shrubs and vines; historical scenes, including a full-scale relief of a typical 1850 canalboat; and even laser projections. The committee emphasized the need for enhancements to be vandal-proof and maintenance-free. The Enhancement Committee provided no specific schedule but suggested that projects could be developed as time and funding allowed.

Of course, eyesores as well as beauty are in the eye of the beholder. A construction engineer or a business owner in the floodplain might consider the floodwall a thing of beauty; a naturalist or a river advocate is likely to hold a different opinion.

The visual impact, however, was not the only riverfront concern raised by the floodwall. Near 17th Street at the beginning of Richmond Dock, which connects the canal to the river, the floodwall would cross the route of the proposed canal restoration. If the floodwall were built as originally designed, the canal would be blocked, severely limiting the prospects for restoration.[176]

One view of the completed floodwall on the northside. *Photo by author.*

The floodwall on southside. *Photo by author.*

"Now the canal enthusiasts were not opposed to the floodwall per se," stated Mary Jane Hogue, president of Historic Richmond Foundation, at the canal dedication in 1999, "but they realized that if provision were not made for the canal to go *through* the floodwall, the opportunity to connect the Haxall Canal near the Tredegar Iron Works all the way to the Great Shiplock in front of Tobacco Row below Church Hill would be lost forever. The canal buffs also felt that a short ride from 12th to 17th streets just wouldn't be enough to thrill the general public or generate economic return."

Since construction of the wall was well underway before this issue was raised, the only viable solution at the time was not a gate but a knockout panel that would preserve the possibility for a gate to be added later. This "change order" was estimated to cost $750,000. Led by Buford Scott, the Canal Committee went back to work raising money and convincing the city administration and property owners along the canal of the importance of the canal passageway. The Corps of Engineers agreed to incorporate the knockout panel if the city would formally request the change and provide the funding for the difference in cost that would be required. To reduce cost, the city requested that the north-side overlook be eliminated and the

148

The very important "last-minute" hole in the wall. *Photo by author.*

funds saved be devoted to the canal gate. "With the funds in hand and the City and the Corps grumbling, the knock-out panel was installed—just in the nick of time."[177]

CULMINATION

"After 22 years of cajoling Congress and local politicians for floodwall money, Dale Wiley finally cracked a bottle of Virginia wine against the giant concrete structure," so proclaimed the *Richmond Times-Dispatch* the day after the dedication of the Richmond Floodwall at 14th and Dock Streets on October 21, 1994. In addition to Dale Wiley, the dedication included a welcome by Mayor Leonidas B. Young and remarks by Dr. John H. Zirschky (acting assistant secretary of the army), Representative Thomas J. Bliley Jr., Representative Robert C. Scott, Senator Charles S. Robb and Senator John W. Warner. The dedication program celebrated the event as well as the history that brought it about:

Twenty-two years ago, the Central Richmond Association Flood Protection Committee was formed to advocate, on a local, state and federal level, for the funding of a floodwall. It took thirteen years of persistence and tenacity on the part of the committee before Congress finally approved the construction of the floodwall. Today's celebration ends a story that began with devastating losses from severe floods and gives the promise of a vigorous future for Richmond's downtown—the heart of the metropolitan area.

A SURPRISE

Almost ten years after the completion of the floodwall, Tropical Storm Gaston arrived in Richmond in 2003. The storm stalled and lingered over the city for hours, dropping twelve inches of rain. Shockoe Slip and Shockoe Bottom, areas inside the floodwall, were inundated, causing significant damage. A disputed claim is that the floodwall prevented the storm's water from draining to the river and thus exacerbated the damage. Since the floodwall was completed in 1994, the river has not risen to a level requiring its use. Of course, that day may come.

PERSPECTIVE

For its advocates, the U.S. Army Corps of Engineers floodwall project is a hard-earned success that has few negatives and is well worth its cost. In his account of Richmond's two worst floods, the flood of 1771 and the 1972 Agnes flood, Walter Griggs concludes that the floodwall in Richmond is "good news."[178] No doubt many other proclamations of the floodwall's benefits could be obtained. After all, it does protect established properties on both sides of the river from flooding and certainly is more desirable than some of the other options considered—like large upstream impoundments or channelization.

Yet from another perspective the wisdom of building the floodwall can be questioned. Was it a short-sighted solution that overvalued a relatively small section of the city and the private property located there? What if there had been a broader, sustained objection to the construction of the floodwall? What might have been the outcome if Ancarrow's position had been heeded and government had let owners take their chances with nature? Or if the initial argument of the Conservation Council had won the day that "in the long run evacuation of buildings and businesses in the flood plain" was the most desirable? What if the cost of the floodwall had been used to purchase these properties and assist owners to relocate? One can wonder what the floodwall-protected portion of downtown would now be like if any of these scenarios had played out.

Such speculation at this point, of course, is not useful. Still, there is little doubt that the floodwall adds a barrier to the visual and physical access to the river. It can be argued that the floodwall makes the James River in Richmond less, not more, park-like. The floodwall fits the premodern paradigm of "engineering the river" better than the modern "let it run free" ethos. And unfortunately, Carlton Abbott's effort to make the wall "disappear" has not been accomplished and Charles Peters's worry about a "permanent eyesore" has come to pass. But to be fair, the ugly riprap-laden south-side floodwall does provide an excellent walking path and a platform for a sweeping view of the city skyline with the river in the foreground.

Perhaps, as the 2012 Riverfront Plan recommends, future actions can ameliorate some of the least attractive features of the wall, especially in the south-side Manchester area. Other suggestions are sometimes offered: paint the tall wall in the Spaghetti Works area to provide "doorways to history" along the river. Murals could depict what the river on the other side of the wall would have looked like in different periods. Or walkways on top of the floodwall could be constructed (all with Army Corps of Engineers permission, of course).

As the construction of the floodwall progressed, so did pursuit of the vision for a vibrant and publicly accessible riverfront in downtown Richmond.

RIVERFRONT

BRINGING THE VISION TO LIFE

Contrary to many riverfront plans advanced over the past 20 years, this plan is real. The property-owners are committed, preliminary engineering and cost studies are complete, and funding has been identified.
—Brenton S. Halsey, 1993

Richmond's Riverfront and river views contribute dramatically to the City's unique sense of place, quality of life, and desirability of property.
—2012 Riverfront Plan

By the late 1980s, the stage was set for on-the-ground riverfront action; six factors had come into alignment. First, a series of plans and proposals, beginning in the 1960s, had built an emerging vision of what the downtown riverfront could be. Second, the city, by purchase and trade, had taken ownership of important riverfront properties, notably Belle Isle and Brown's Island. Third, interest in the river by paddlers, other recreationists and general river enthusiasts had grown as part of a national trend and as a result of a cleaner river. Fourth, through the Richmond Renaissance Discover the James Program and the organization of major riverfront businesses into the Richmond Riverfront Development Corporation (RRDC), an instrument for coordination and cooperation with the city had been established. Fifth, the floodwall project was underway, bringing attention to the riverfront along with the realization that it must be "done right" or options would be foreclosed permanently. Finally, the

EPA mandated that a combined sewer overflow (CSO) interceptor had to be placed somewhere along the downtown riverfront, providing an opportunity for canal and riverfront restoration to piggyback on a funded project. This combination of factors provided the basis for action. Fundamentally, two interests sometimes in conflict—public recreational use of the river and economic interests of private landowners—were aligned for cooperation.

By the time the floodwall was completed in 1994, restoring the riverfront had already begun, and the city and a group of riverfront landowners had reached agreement for an even more ambitious project.

RICHMOND RENAISSANCE: GETTING STARTED ON RIVERFRONT DEVELOPMENT

Richmond Renaissance, a public-private corporation formed in 1982 to stimulate economic development in downtown Richmond, also joined the riverfront renewal effort. Renaissance had the advantage of a membership made up of major property owners and top city officials, an indication that the venture was a serious one. The first priority of Renaissance was Jackson Ward and the Sixth Street Marketplace, but the riverfront became a close second and, some would say, a more successful venture. The James River portion of Renaissance was announced with enthusiasm: "Scenic in its beauty, historic in its character, the James has been shut off from public access and use for decades. A $4,200,000 James River Discovery Program is proposed with the specific goal of opening the James River so that residents of the city and the metropolitan area can begin to enjoy this natural resource and understand its vast potential."[179] In addition to public resources, corporate owners in coordination with the Renaissance Riverfront Committee committed funds to improve their riverfront properties.

Groundbreaking ceremonies for the James River Discovery Program were held in July 1986 on Brown's Island. Directed by Marc Hirth, that program included an early canal walk on the Haxall Mill Race, improvements to Brown's Island and Tredegar Street, as well as Belle Isle improvements. At the time, Belle Isle was inaccessible from the north shore. Over the years, several plans had been proposed for providing access, including a ferry and a monorail in Stanley Abbott's early park plan. More recently, a bridge on top of old railroad bridge piers was considered, but that gave way to a new idea—a cable-supported pedestrian bridge hung

A view of the cable-supported pedestrian walkway under Lee Bridge alongside the remnants of an earlier bridge to Belle Isle. *Courtesy of Scott Adams.*

under the new Lee Bridge, which was under construction at the time. Halsey calls the work on Belle Isle and the pedestrian bridge completed in 1991 the "piece de resistance of the Discovery project."[180] Belle Isle has become the most visited location in the James River Park System, which in turn is the most visited entity in Richmond.

CANAL WALK

James River Discovery cleared the way for more extensive riverfront improvements to follow with canals still at the centerpiece of that effort. Its action opened the downtown riverfront to recreation and helped demonstrate the potential the riverfront held, as one reporter put it, to be both "scenic and lucrative."[181] Having begun the downtown riverfront transformation and foreseeing the complexity that lay ahead, Renaissance handed off the development effort to a more specialized public-private partnership—the newly created Richmond Riverfront Development Corporation (RRDC) made up of riverfront landowners and the city. The businesses included in the RRDC, Ethyl Corporation, Dominion Resources, Norfolk Southern, Reynolds Metals and the Ladybyrd Hat Building Partnership, agreed to donate property, honor restrictive covenants and contribute funding while the city committed to major infrastructure improvements in the area.

"Key to the progress," as Brenton S. Halsey, president of the RRDC, framed it in an op-ed in 1993 on the occasion of plan unveiling, "has been the creation of the Richmond Riverfront Development Corporation, the brainchild of the City Manager. Following its practice of catalyzing economic development ideas in their infancy and turning them over to other appropriate organizations for implementation, Richmond Renaissance has transferred its efforts into the new development corporation. The corporation is a joint venture of the city and the five major landowners in the development area. It is a true public/private partnership and is a unique development tool unlike any in Virginia."[182]

The central document used in moving forward was the 1993 *Richmond Riverfront Master Development Plan* prepared by Wallace, Roberts & Todd (architects and planners) and Greeley and Hansen (civil and hydraulics engineers). This plan, which was built on the 1988 and 1991 Abbott canal restoration plans, combined canal restoration and CSO interceptor installation in the canal bed. And, to quote Halsey again, "Contrary to many riverfront plans advanced over the past 20 years, this plan is *real*. The property-owners are committed, preliminary engineering and cost studies are complete, and funding has been identified."[183] Action did follow.

The decision to place the EPA-mandated combined sewer overflow interceptor beneath the canal bed rather than the river as originally planned was a critical financial boost in enabling the canal restoration to go forward. This idea, credited by Halsey to Robert Bobb, city manager, reduced the cost of both projects. Further, as the Falls of the James Scenic River Advisory Committee argued in a letter of support to the Virginia Marine Resources Commission, the canal location avoided damage that likely would have been done to the bank of the river.[184]

The challenge tackled by the Richmond Riverfront Development Corporation, was to develop thirty-five acres of downtown property "for a high grade mix of new construction and rehabilitation including offices, retail stores, restaurants, and residential uses." The vision is "a continuous public recreation experience extending from Belle Isle and Valentine Riverside [Tredegar] to the Great Ship Lock." This plan laid out the course for a reconstructed canal and canal walk and provided detailed architectural and landscape guidelines. It included the decision to join the CSO interceptor

Canal restoration in progress. *Courtesy of Venture Richmond.*

project and the canal reconstruction both as a physical reality and a funding mechanism. The plan was considered the first phase for achieving the "ultimate objective of continuous canal redevelopment, from Great Ship Lock, west to Maymont Park." The aspiration, much like earlier plans, is no less than to "change the face of Richmond."[185]

A trip to San Antonio, Texas, by city and RRDC officials and businessmen to investigate the renowned Paseo del Rio or River Walk was an important event for both inspiration and ideas. The River Walk, "brimming with stores, restaurants, nightclubs, dinner boats and water taxis,"[186] was considered the "best existing model for the Richmond riverfront."[187]

The first phase called for canal restoration with a continuous walkway from a point near Tredegar to 17th Street, a distance of more than a mile. Since most of the original Kanawha Canal was no longer useable in this stretch, the Haxall (formerly a millrace for water powered manufacturing) was substituted. This left a gap from the end of Haxall to the west end of Richmond Dock (the final stretch of the Kanawha to the river). A new canal and turning basin was constructed to fill this gap, no small feat since the area was cluttered with streets, railways, underground sewage and drainage pipes and the piers of I-95. Two critical points in design were the connection between Haxall and the newly constructed route, a location of significant elevation change, and the entry into Richmond Dock reached through the floodwall opening, which had been added at the "last minute" in floodwall construction (see chapter 8).

History was to be a distinctive feature of the Richmond canal project. In 1997, the RRDC appointed a Historic Interpretation Committee; after an RFP process, it hired Ralph Applebaum & Associates of New York to design an outdoor museum along the restored canal. Richmond Historic Riverfront Foundation was formed to handle donations and finances. Much of Applebaum's plan, "A Vision for Richmond's Historic Riverfront," was brought to life with a series of exhibits, historic medallions in the walkway, floodwall murals and interpretive and orientation pylons (see chapter 11).

Not long before the restored canal's eagerly awaited opening, the plan for tourist boats hit an obstacle—a legal-bureaucratic one. The U.S. Coast Guard, based on historic use, declared the James River and Kanawha Canal a navigable waterway, thus subject to its regulations.[188] Coast Guard requirements would have meant expensive hull design and extensive operator testing, perhaps costly enough to scuttle canalboat plans. Coming

The new turning basin along Canal Walk, not in the same location as the original Great Turning Basin, which is now under high-rise buildings. *Photo by author.*

to the rescue, Representatives Thomas J. Bliley Jr. and Robert C. Scott coauthored legislation declaring the canal nonnavigable. Objections were raised by the House Transportation and Infrastructure Committee, but after several weeks of lobbying, a visit to the canal by congressional staff and an ordinance introduced by Mayor Timothy Kaine, compromise was reached and the three-foot-deep canal was declared legally nonnavigable.[189] (It should be noted that, if it ever occurs, making the Great Ship Lock operational, a long-standing dream of many canal enthusiasts, would again raise the issue of the navigable status of the canal.)

In 1999, the new downtown riverfront, now called Canal Walk, was ready for unveiling. Formal dedication of the restored canals and Canal Walk took place on June 4, 1999, with the presence of an enthusiastic crowd and great assortment of dignitaries, including three former governors: Robb, Baliles and Wilder. Chair of the Richmond Riverfront Development Corporation Brenton Halsey presided, and Mayor Timothy Kaine delivered the keynote speech. A gala fundraising celebration was held later on the evening of June 12.[190]

Brown's Island after completion of Canal Walk—a very different place than the industrial use of decades before. *Photo by author.*

When the restored canal and Canal Walk opened in 1999, the long-standing vision of making the James River the centerpiece of the city was closer than it had been in the modern era. Not only was the river more accessible than it had been since industrial days, but the city also was taking symbolic action to recognize this new emerging aspiration. Also in 1999, "George [Washington] on a horse" was retired, and the city logo was redesigned with the James River as its centerpiece. So at least symbolically, the goal had been achieved. But there was still a long way to go. The canal had been restored, the walkway constructed and historic signage and memorabilia installed or preserved, but the businesses, shops and restaurants were yet to arrive.

DEVELOPMENT ALONG CANAL WALK: MEETING ECONOMIC OBJECTIVES

Economic development was a key objective for the restored canal and Canal Walk undertaking, and development would be the measure of its success. Success would require a canal lined with hotels, restaurants and residences, all of which would take time. Leaders of the planning and development effort offered great expectations about the future the new riverfront would stimulate. "If all goes according to expectations," said Mark Hirth, executive director of the Richmond Riverfront Development Corporation, "the project will generate private development worth $450 million, which will result in $10 million a year in taxes."[191] Tourism was expected to generate "$100 million during the first 10 years."[192] Governor George Allen proclaimed the riverfront development to be "chock full of potential."[193] Hopes were expressed in nonmonetary terms too: "I'll know the canal is a success when I can order a margarita at a sidewalk café and watch the boats go by" said Jack Berry, executive director of Richmond Renaissance.[194]

Still, even before the opening of the Canal Walk, cautionary tales were floated—don't expect development to occur instantly. It will take time. "Optimism runs high; realism is urged" was a caption in the *Times-Dispatch*. To invoke a cliché, cautious optimism was the order of the day. Some feared that the Canal Walk would have the same fate as other Richmond projects—namely the Sixth Street Marketplace, Valentine Riverside and Main Street Station. Each had opened with great fanfare and then fizzled. In the planning and construction phases for Canal Walk, the San Antonio River Walk had been held up as the model being emulated and as an indication of the economic success that could be expected. Following the cautious optimism theme, developer Mark R. Merhige quipped, "You don't just add water and have San Antonio."[195] San Antonio took thirty years to become the success for which it is now recognized.

The first few years after the Richmond Canal Walk opened, it had the feel of a ghost town except on those occasions when special events were held. It seemed to be a "chicken or egg" problem. Until businesses offering amenities were there, the people would not come; but until the people came, amenity businesses were unwilling to locate there. As Halsey, a prime mover in the enterprise, warned, "It's going to take at least five or ten years to determine whether or not it's really a success." Without development, the Canal Walk will be "nothing more than a pretty park."[196]

More than fifteen years after dedication of the Canal Walk, in his 2016 memorialization, *Riverfront Renaissance*, Halsey concludes that downtown canal development has been both successful and disappointing. He estimates that new investments in the restored area range from $462 million to $650 million, depending on the size of the area included. This compares quite favorably to the hopes expressed as the riverfront plans were being unveiled. And investment opportunities are available, so the numbers may grow. On the other hand, the ambience of San Antonio's busy canal front restaurant and shop scene has not been achieved.

One drawback the Richmond Canal Walk faces, in comparison to San Antonio's River Walk, is the inability of canalboats to travel the full length of the restored canal. Although there is a continuous pedestrian walkway, albeit with a formidable set of stairs, boats cannot travel the full length of the restored canal because of elevation change. Options were available, and still are, to make this connection navigable with the use of the preserved locks and a connector channel, but the cost of making this connection boat-worthy was considered prohibitive and thus not pursued. Consequently, it is not possible to use a canalboat to get from one end to the other or for a canalboat to serve as a taxi to a restaurant on the opposite segment of the canal. Money and the required will could change that according to Halsey, who has laid out a strategy for connectivity, most of it by water, from Bosher's Dam to Rocketts Landing.[197]

The restored canal is considered by many to be the "new riverfront." While the restored canal and Canal Walk do provide a link to the river superior to the conditions that preceded, a direct connection to the river is offered in only a small portion of the total walkway. In other segments, the river is hidden by the floodwall or the Shockoe Retention Basin. Although

The Canal Walk in spring. *"Cherry Blossom Time in Richmond" by Nancy Helms, courtesy of Scenic Virginia.*

not part of the James River Park System, the Canal Walk is a park in its own right and comes to life when special events take place. Building Canal Walk and associated riverfront improvements, despite the shortcomings a critical eye might identify, is a major accomplishment. It has and likely will continue to attract investment, it is an outdoor museum, it does host multiple events with large crowds and it is an anchor and stimulus for further riverfront enhancement and even further canal restoration. It is a beginning and not an end.

AFTER CANAL WALK: MORE TO DO

For the next decade, ideas surfaced, controversies erupted, officeholders were lobbied and the potential for the downtown riverfront, still not fully achieved, was kept alive. Frequent reference was made to the river and areas that might be improved, developed or protected. Height restrictions on riverfront development were considered and adopted by city council (see chapter 10). New residential and commercial buildings were constructed alongside the restored canal and other important projects, like the north-side trail and a downtown whitewater boating takeout were completed. Nevertheless, there was much unfulfilled potential—the Kanawha Canal to Maymont had not been restored, nor the Byrd Park Pump House adequately preserved, nor the trail on Chapel Island opened, nor Mayo's Island made accessible to the public, nor the old docks area downriver restored, nor much of anything done on the Manchester side of downtown.

The process for revising the Downtown Master Plan helped keep the riverfront on the political agenda, and the proceedings brought out ideas for riverfront improvements as well as disagreements about how the riverfront should be used. As framed in a report on a *Richmond Times-Dispatch* Public Square, the central challenge is "charting a course between economic development and preservation of the river's scenic and natural gifts."[198] Giving focus to the larger issue and a continuing source of disagreement were two riverfront properties under private ownership that many hope will be devoted to public purpose: Mayo's Island in the center of downtown and the USP/Echo Harbour property adjacent and downstream from Great Ship Lock. Mayo's Island, crossed at grade level by Mayo's (aka 14th Street) Bridge, is visualized by some as a public park, as is USP/Echo Harbour, which is periodically proposed as the site for a high-rise riverside development. Echo/Harbour, if developed as sometimes proposed, would not only mean the loss of a desirable park site but also would mar the iconic view from Libby Hill Park. How to treat these two properties delayed the approval of the Downtown Master Plan revisions more than once.

After those delays, though, the amendments to the Downtown Master Plan were approved by city council on July 27, 2009, with a 9–0 vote. In that same meeting, another action was taken that moved riverfront planning forward. Council approved an ordinance authorizing city purchase of Lehigh Cement property, beginning a process that opened up new public riverside opportunities.[199] The Lehigh property on the riverbank was dominated by tall cement silos that obscured the Church Hill and Libby Hill views.

2012 RIVERFRONT PLAN: CONNECTIVITY

The Downtown Plan revisions sparked a new round of activity and investment in a focused and comprehensive riverfront plan. Hargreaves and Associates was hired and began a riverfront planning process perhaps more ambitious than any that had preceded it. Beginning with extensive citizen involvement, the Hargreaves riverfront planning process covered the river in downtown Richmond, essentially from the Lee Bridge and Belle Isle downstream to Rocketts Landing and included both sides of the river. Past plans had seldom given much attention to Manchester and the south side.

The resulting 2012 Riverfront Plan presented new ideas and also reflected many of the aspirations that had gone before. As described by Mark Olinger, Richmond's director of planning and development review, the new plan, as did several that went before, aspired to "rethinking the city's relationship to the river." "Connectivity" was the overriding theme: "one river, one city, one riverfront system." It was designed to provide "more access to more people more of the time" and "to create a single, unified, cohesive system."[200]

Creating connectivity on the downtown riverfront in Richmond, a landscape that has been for many years disjointed and disconnected, is no small task. Movement, whether pedestrian or vehicular, and whether toward the river or along it, is interrupted by the accumulation of historical decisions: private property lines, floodwall segments, rail lines, streets, city utilities and even a restored canal divided into two segments. Clutter rather than continuity typifies the riverfront the planners had to work with. The plan made great strides toward the connectivity goal with virtually every foot of the riverfront from Belle Isle to Rocketts Landing included. Connectivity was the priority, reflected in two early projects: the completion of the Virginia Capital Trail to downtown Richmond and the Brown's Island Dam Walk. Importantly, both sides of the river were included in the plan.

VIRGINIA CAPITAL TRAIL

Connectivity on a grand scale was achieved with the 2015 completion of the Virginia Capital Trail, a fifty-two-mile paved cycling-pedestrian path between Richmond and Williamsburg/Jamestown that had been underway for a number of years. This major project, a public-private partnership between the Virginia Department of Transportation and the Virginia

Capital Trail Foundation, dovetailed nicely with the city's riverfront plan. The trail runs from the eastern end of Canal Walk at 17th Street along the length of Richmond Dock, an extension of the Kanawha Canal, to Great Ship Lock. A costly aspect of this segment of the trail was the construction of the protective barrier under the CSX rail trestle, which runs adjacent to the canal and above the trail. From Great Ship Lock, the trail continues past Intermediate Terminal, Rocketts Landing and then another fifty miles to Jamestown.

BROWN'S ISLAND DAM WALK

The Brown's Island Dam Walk, later officially named the T. Tyler Potterfield Memorial Bridge after the unexpected death of its principal planner, was selected as the first major project for implementation in the 2012 Riverfront Plan. This project (often respectfully referred to as T-Pott) exemplified connectivity—it connected the north bank with the south, two parts of the city that had been economically, socially and symbolically separated for decades. Further, it satisfied an important principle of plan implementation: begin with a project that can be completed relatively quickly, will be visible and will generate support that will carry over to future projects.

The dam walk was to be built on the Brown's Island Dam, which was constructed in the early 1900s to divert water into the Haxall Canal to produce hydroelectric power. Retrofitting the old dam as a walkway across the river had been suggested many times by outdoor groups and others, but it was usually considered too expensive and impractical. This dam contained sluice gates that could be opened and closed and included a broad walkway on top for workers to manage those gates. It virtually recommended itself as a river-crossing walkway. An overlook that extended some fifty yards into the river had been constructed on the dam during the development of Canal Walk and Brown's Island, but that stopped well short of crossing the river. Further, this overlook, imbedded with quotes from those who had witnessed the fall of Richmond in 1865, was disconnected from the remainder of the dam by a purposely constructed gap to protect adventurous, and sometimes inebriated, persons from the risks of a midnight stroll on the old structure.

After the Riverfront Plan was rolled out, great expectations, sometimes expressed with near poetic words, accompanied the prospect of the Brown's Island Dam being retrofitted as a dedicated pedestrian-cycling crossing of

The T. Tyler Potterfield Memorial Bridge heading south. Although a long time in coming, it became an instant hit. *Photo by author.*

the James, just seventeen feet above the rapids. "Imagine walking a direct line from the north bank to the south, just mere feet above the rushing James. There is no pressure to hurry across and no need to be hyper-vigilant as you make your way. In fact, there are several spots along the span of the Dam Walk that invite you to stop and stare out on the water, and just be there."[201]

With city and state funding, planner persistence and nudging by river advocates, the dam walk was completed in late 2016, although the hope of completion before the 2015 UCI Road World Championships was not to be. The lack of bids in response to the city's request for proposals and the restriction on work during the spring shad spawning season delayed the project, but on December 2, 2016, Mayor Dwight C. Jones, with a large crowd in attendance, cut the ribbon dedicating the accomplishment.

The T. Tyler Potterfield Memorial Bridge (T-Pott) has lived up to and even exceeded expectations. The walkway has been a major draw to the river, attracting visitors who had seldom if ever ventured to the downtown riverfront. Thirty-five thousand visits were counted in its honeymoon month. The success of the dam has been underlined by an American Public Works Association award naming it one of the 2017 "public works projects of the year."[202]

BRIDGE PARK

About the time the 2012 Riverfront Plan was being finalized and approved, a new idea with significant implications for the downtown riverfront surfaced from an unexpected source. Ella Kelley and Mike Hughes of the Martin Agency began to promote the idea of a "bridge park." Their initial idea was to retain the old Huguenot Bridge, well upstream from downtown, when the new bridge was completed and convert it to park space, but their proposal came too late to receive serious consideration. With the Huguenot Bridge location unavailable, Kelley and Hughes sought a new venue and ratcheted up their efforts.

The new proposed location is downtown Richmond with the still surviving piers of the old Richmond-Petersburg Railway Bridge as the foundation for Bridge Park. That idea has gained momentum and is now being pursued by a nonprofit 501(c)(3) foundation—Bridge Park Richmond. Ted Elmore, president of the foundation, is leading a design, community consultation, fundraising effort. Multiple designs are being considered, some of which would and some that would not use the railway piers. With Potterfield Bridge a reality, a question is whether Bridge Park would be redundant or a complement. The proponents contend that the over-water park would be a "destination" that would enhance rather than compete with the Potterfield Bridge. Certainly, the two together would make an exceptional pedestrian over water loop in downtown Richmond. Not part of the riverfront plan and with no public funding, Bridge Park is still in the concept stage.

TREDEGAR GREEN

Even though suggested frequently over the years, Potterfield Bridge likely would not have come about without the impetus of the 2012 Riverfront Plan. Tredegar Green, a natural amphitheater on Gamble's Hill near the Lee Bridge and the entrance to Belle Isle, was envisioned in earlier plans as well as the Riverfront Plan, but it might well have been successfully pursued without endorsement of the 2012 plan. The prime mover for Tredegar Green was Venture Richmond, the successor to Richmond Renaissance and the Richmond Riverfront Development Corporation. Venture Richmond, led by Jack Berry at the time, wanted a new and expanded location for the popular Richmond Folk Festival held every fall.

Unlike the Potterfield Bridge, which proceeded from plan to implementation with little but enthusiastic support, the Tredegar Green proposal stirred considerable opposition from two sources: residents of nearby Oregon Hill who were concerned about noise and traffic congestion and canal preservationists who feared damage to the dry canal bed that crossed the four-and-a-half-acre site. The Tredegar Green proposal included narrowing the bed of the canal and lowering the towpath to provide a better sightline to the stage at the bottom of the hill. The concern of the canal advocates was not simply damage to the long-abandoned canal but action that would preempt the future possibility of resurrecting the canal so that boats could run from Tredegar to Maymont and the Pump House.

Modifications to the amphitheater plan were made and approval received. Not all were satisfied, but the worst fears in each camp were mollified. Proponents received approval to proceed, although with some conditions, and opponents were assured that future watering of the canal would remain possible.

DOWNRIVER PLAN

As Potterfield Bridge was nearing completion, the second major Riverfront Plan implementation task got underway—the downriver section from Great Ship Lock to Rocketts Landing at the Henrico County line, including the Lehigh Cement property, Intermediate Terminal and the old Richmond port area. The riverfront in this tidal portion of the James had been an industrial/shipping center (as well as home of the Confederate Navy Yard) in the period before commercial ships became too large to navigate its waters.

This downriver undertaking was moved up from Priority 2, bypassing such projects as Missing Link Trail and the acquisition of Mayo's Island, both expensive and difficult projects. Three developments had set the stage for riverfront renewal: the city's purchase of the Lehigh Cement Company property and the subsequent removal of the cement silos, the agreement with Stone Brewing Company to lease the Intermediate Terminal for a restaurant and beer garden and completion of the city's portion of the Virginia Capital Trail. The only piece missing in the stretch from Great Ship Lock to Rocketts Landing was the USP/Echo Harbour property, which was still in limbo—privately owned, but with no approved plan for development or sale.

Hargreaves and Associates, with the Richmond Department of Planning and Development Review, undertook the task of fleshing out this downriver portion of the 2012 Riverfront Plan. Alternative concepts were developed, multiple public feedback sessions held in city hall and elsewhere and various river-interested groups consulted. Planners had a variety of issues and conundrums to address in developing a detailed approach for these five acres of old industrial riverfront:

- Design while recognizing that the entire area is in a flood zone.
- Determine how to provide access for multiple forms of mobility—driving, biking, walking, boating. Alternatives to driving were especially important given the difficulty in providing parking.
- Determine what should be greenspace and what hardscape and how to maintain and restore natural features like trees for shade, natural riverbanks and river views.
- Determine how to provide for a full range of users, including fishermen who were regulars on this portion of the river.
- Determine how to recognize and interpret the history of the area.
- Determine what kind of recreation amenities to provide. River advocates argued for river-related themes like splash pads rather than traditional playground equipment.
- Determine whether this riverfront area should offer kayak, canoe and SUP access from land and, if so, how.
- Restore the dock to accommodate small cruise ships.
- Plan so the Virginia Capital Trail, which bisects the area, will be an asset and not interfere with other uses.
- Determine, with Stone Brewing, the relationship of the restaurant/beer garden with public space.
- Determine how to provide for public events and viewing of special events like Parade of Lights and Dragon Boat Races.
- Determine where to locate restroom facilities.
- Structure the layout so efficient maintenance and trash collection can be facilitated.

Despite the challenges, an approach, with many details still to be determined, was approved in 2017 by the planning commission and city council as a Riverfront Plan Amendment to the City's Master Plan. The schedule for more detailed planning, funding and full implementation is uncertain.

PERSPECTIVE

The restoration/construction of the downtown riverfront has been and continues to be a complex puzzle, fitting together pieces that sometimes mesh and sometimes do not. Private property, commerce, recreation, flood protection, historic preservation, public utilities and transportation each has its constituency as well as physical presence on the ground. Sometimes two or more of these sectors are mutually reinforcing; at other times, they are in conflict. Building the Downtown Expressway and preserving the Kanawha Canal and Tidewater Connection Locks were in conflict, and something had to give. Building the combined sewer overflow receptor and restoring the Haxall Canal and its extension were not only compatible, but the canal project might never have been done without the funding provided by the federally mandated interceptor.

Whether compatible or conflicting, the decisions were complex. A tabula rasa it definitely was (and is) not. Planners, advocates and decision makers were working with a cluttered landscape with infrastructure both under- and above ground. Several levels of government were involved along with multiple property owners, both public and private, and a collage of advocacy groups and stakeholders. Conflicts were not always just about what should happen along the river; the city always faced other demands for its attention and resources. At times, river advocates have been frustrated with city government for its seeming lack of interest in the river and its attention to competing matters. At other times the city has led riverfront efforts. Whether led by citizen activists, business leaders or the city government, a great deal has been accomplished.

There is no doubt that the city is more oriented toward the river now than it was sixty or seventy years ago; even the city's logo has been changed to include the river as a feature. The downtown riverfront is a big part of that orientation. From Belle Isle and the hanging pedestrian walkway under the Lee Bridge to Rocketts Landing in Henrico County, the north bank of the river is far different than it was when an early plan was published in the 1960s. Parts of the riverfront have been preserved or restored and other parts altered, all hopefully for the better.

Perhaps just as important as the on-the-ground accomplishments, the City of Richmond has adopted and begun to implement a riverfront plan that should guide action for years to come. Fundamentally, the river is valued as it was not in earlier decades. Although plans can outpace implementation, the extent to which today's downtown riverfront matches the vision displayed in the plans of the 1960s, 1970s and 1980s is cause for optimism. The 2012 Riverfront Plan and other ideas of river enthusiasts make clear that more can be done.

———◦◦◦———

Enjoying views of the James is a centuries-old recreational activity. Concern about protecting some of these river views became a salient public issue shortly after dedication of the Canal Walk.

10

VIEWS

THE RIVER AS AESTHETIC EXPERIENCE

The curve of the James River and steep slope on this side are very much like the features of the River Thames in England, at a royal village West of London called Richmond upon Thames.
—plaque at Libby Hill Park

These priority views are public amenities. They have significant public value and/or historic importance and should be defined, preserved and enhanced for the community.
—2012 Riverfront Plan

The views they offer are a valued feature of most rivers. Artists, poets, essayists and photographers extoll rivers' beauty and the inspiration they provide—and the James in Richmond is no exception. Many views of the James River were celebrated in centuries past by residents and travelers alike, as "striking and beautiful," as "a magnificent picture."[203] Some of these striking and beautiful vistas remain, but many have been blocked or cluttered by development. At one time, the portico of the Thomas Jefferson–designed Virginia Capitol provided a view of the river "so directly underneath you that it almost seems that you could leap into it."[204] That view no longer extends to the river.

Other views, too, have disappeared, but many remain and some have returned. During the heyday of water-powered industry, the banks were lined with factories, making the river all but invisible from the remainder of the city. This visual wall has been taken down by a changing economy and only partially replaced by more modern structures, one of which is the floodwall. So, the dance between destroying and protecting public views continues. Some dramatic views, including some of historic significance, are endangered. Protecting, or even enhancing, these public river views is a struggle some individuals and organizations have taken on. Three cases in particular offer a sample of the issues raised and the passions stirred as developers and view protectors clash: the proposed expansion of Dominion Resources headquarters on Tredegar Street, change in riverfront zoning ordinances and efforts to protect the view from Libby Hill Park.

DOMINION RESOURCES VERSUS CITIZENS ADVOCATING RESPONSIBLE RIVERFRONT DEVELOPMENT

Dominion Resources (now Dominion Energy) occupies the property at the westernmost end of Tredegar Street as its headquarters. This location, at the foot of Oregon Hill, provides a dramatic close-up view of the river and its rapids from Dominion's grounds and many of its offices and conference rooms.

In 2001, Dominion sought city approval to expand the buildings at this site to create office space and parking so that it could relocate nine hundred or so jobs from its Innsbruck office to the offices here. The expansion would potentially raise the height of these buildings along the banks of the James to a level that could partially obstruct the view of the river from the Oregon Hill bluff, which holds a city park and gentrifying residential area. The effort set off several months of struggle between development and view protection.

To pursue its building plan, Dominion requested rezoning to B-4, which would allow building to an unlimited height. The plans would include an office tower of 175 feet as well as a separate energy trading center and a parking deck. Dominion argued that its expansion would bring jobs, economic growth and new tax revenues to the city. Dominion's rezoning request did not go unchallenged.

Opponents argued that expansion in the city is applauded, just not at the riverside Tredegar Street location. The most vocal opponent of the view-blocking expansion was Citizens Advocating Responsible Riverfront

The view from a Dominion Energy conference room generously provided by Dominion for a meeting of the James River Advisory Council. *Photo by author.*

One of the views from Oregon Hill looking over Dominion Energy office buildings. *Photo by author.*

Development (CARRD), a group initially composed of Oregon Hill residents but expanded to include many other organizations and individuals. Activity around this rezoning request heated up in the summer and fall of 2001. CARRD recruited members and supporters and lobbied vigorously against the B-4 zoning. Dominion plans, it contended, "would degrade the views from Oregon Hill Park, Hollywood Cemetery, Belle Isle, Woodland Heights, the Lee Bridge, and the river itself."[205]

At meetings and hearings and through emails, presentations and letters to the editor, CARRD argued against Dominion's plan with the hope of building enough opposition that Dominion would alter its plan or, failing that, dissuade public officials on the planning commission and city council from approving the rezoning request. Both Dominion and CARRD wanted to use the beautiful setting the Tredegar location offers—Dominion to afford the setting to more of its employees and visitors, CARRD to preserve the view for the residents of Oregon Hill and the public.

PLAN MODIFICATIONS—AN EFFORT TO MEET OBJECTIONS

Dominion and its representatives heard the objections to its plans many times in meetings with the community and in other forums. Part of Dominion's response was a series of modifications to the plan itself. In a press release on September 6, 2001, Dominion stated, "Neighborhood views of the historic James River will be preserved under newly revised plans for a campus expansion by Dominion at its riverfront headquarters on Tredegar Street in downtown Richmond." The press release was reporting a "Declaration of Restrictive Covenants" that had been filed the day before in Richmond Circuit Court. This declaration included several height restrictions but had an important proviso: "provided that the B-4 zoning is granted and not subject to appeal."

The press release framed the declaration in terms intended to soften the objections of opponents: "Dominion's filing shows that the height of the company's planned energy trading center would be no more than 65 feet high. This is 45 feet shorter than original estimates proposed by Dominion earlier this year at neighborhood meetings. Those meetings were initiated by the company to solicit neighborhood opinion as it developed plans for expansion."

The release continued: "The filing also shows that the new office building would be no taller than 175 feet in height—or 65 feet taller than the existing campus high point of 110 feet. Moreover, the new building would be located behind the existing 110-foot office building. By concentrating the height in one area of the site, the impact on views from adjoining properties is minimized."

Dominion project director, Wes Keck, argued that a "win-win solution" had been found. "These height restrictions and our planned architectural design can balance Dominion's expansion needs and Richmond's economic development needs with the neighbors' desires to keep the view of the river."[206] In addition to statements claiming that new height limitations and carefully constructed architectural designs were responsive to the concerns of opponents, the release included quotations from James W. Dunn, president and CEO of the Greater Richmond Chamber of Commerce, and Gregory H. Wingfield, president and CEO of Greater Richmond Partnership Inc., touting the high-paying jobs and increased tax base the plan would add to Richmond's economy.

Opponents of Dominion's plan were not persuaded by the modifications and continued their energetic opposition with a multi-pronged argument. "The viewshed from Oregon Hill Park overlooking the James provides one of the best public river views in Richmond. It is a stunning, irreplaceable, and unique natural resource, revealing the rocky sweep of the James facing west and the natural beauty of James River Park and Belle Isle facing south."[207] The Falls of the James Scenic River Advisory Committee took an active role in the dispute, largely as a sounding board and an intermediary between CARRD and Dominion Resources, which was represented by the McGuire Woods law firm. The Falls Committee held multiple meetings in which each side made its case. While endeavoring not to be "anti-development," the committee did seek to protect the river viewshed.

Dominion had vocal opposition, but it also had its supporters. In addition to the chamber of commerce and the Greater Richmond Partnership quoted in the press release, a number of prominent voices, including the *Richmond Times-Dispatch*, encouraged approval of Dominion's plan. "What's more," an editorial stated, "the current controversy delays an influx of more than a thousand workers and $66 million in payroll to downtown. That buys a lot of lunches at area bistros. The Planning Commission and the City Council should approve Dominion's rezoning request without further delay."[208]

NEXT STEP—A CHANGE IN TACTICS

By unanimous vote on October 1, the planning commission rejected Dominion's application for rezoning to B-4, which brought about a shift in tactics for Dominion. The planning commission recommended that Dominion shift the rezoning request to B-5, which allows sixty-foot height, and seek a special use permit to go higher. The commission reasoning was twofold. The special use permit would allow the city greater control than B-4 zoning. Further, B-5 would not set a bad precedent for future riverfront building, as some commissioners thought that B-4 would do. The commission did not indicate a height it would or would not approve but left that for Dominion to work out with the neighborhood.[209]

Dominion's response to the planning commission "invitation" was rapid. Four days after the commission meeting, Dominion submitted a new rezoning application, this time asking for B-5 zoning and a special use permit for a 160-foot building, 15 feet lower than its earlier request.

The new request received partial support from opponents—support for the B-5 zoning but adamant opposition to the special use permit for a 160-foot building. The weeks that followed continued to see a contest waged between opponents and supporters of Dominion's newest plan in the form of protests, letters to the editor, testimony at meetings, editorials in the *Richmond Times-Dispatch* and *Style Weekly* and full-page ads in the *Dispatch* by Dominion in support and by Scenic Virginia in opposition.

In addition to the 160-foot building, attention was also given to the parking deck expansion and proposed trading floor building, arguing that those buildings too would block river views from Oregon Hill Park unless kept lower than proposed and/or sloped toward the river. The Falls of the James Scenic River Advisory Committee made several site visits to the park to visualize the extent of obstruction. In presentations to the planning commission and city council, committee chair Dr. R.B. Young consistently argued that these buildings would block more of the view than desirable and should be modified. He recommended sloping the roof toward the river, which would enable a fuller view from the bluff.

In the midst of vigorous last-minute lobbying by both sides, the stage was set for a recommendation by the planning commission and a decision by city council. At the November 5 meeting of the planning commission, Gloria Freye of McGuire Woods made the applicant presentation on behalf of Dominion, followed by a passionate public hearing. Dominion supporters included Richmond Renaissance, the Greater Richmond Partnership, the Metropolitan Business League, the Greater Richmond Chamber of Commerce, Richmond Riverfront Corporation and Richmond Marriott. Several members of CARRD were among the opposition who spoke, as well as Scenic Virginia, Falls of the James Scenic River Advisory Committee, the Sierra Club and several individuals. The planning commission endorsed the application for forwarding to council in a 5–2 vote. Two commissioners left the meeting before the vote.[210]

CARRD and other opponents continued their efforts as the Dominion application made its way to council. The day of the city council meeting, November 26, Scenic Virginia placed a full-page ad in the *Richmond Times-Dispatch* that likely captured the sentiments of most opponents:

> *Rezoning the site to B-5 has overwhelming citizen support. City Council should and certainly will approve it. Both hard logic and citizen sentiment, on the other hand, overwhelmingly oppose the special use permit. What Dominion proposes would do irreparable damage to the city's parks and quality of life, its tourist appeal, and the city's master plan of development....But they* [council members] *are finding it hard to say no to Dominion. You can scarcely overestimate the pressure that a corporation with Dominion's vast resources—with a budget that dwarfs Richmond's own—can bring to bear.*[211]

City council heard many of the same speakers at its public hearing as the planning commission had heard earlier, with some additions, including a woman dressed as the devil who called herself "R. Lucy Furr."[212]

City council voted unanimously to accept the rezoning application and the special use permit for a 160-foot office building. As *Style Weekly* reported, "Droves of protesters mourned and went home" and an "even greater army of supporters...cheered and went home"[213] The fight had boiled down to dollars versus views—and dollars won.

POSTSCRIPT

Dominion did build its trading floor, and its roof does slope toward the river, expanding the view it allows. To date, more than eighteen years later, the 160-foot office building that was the most controversial has not been constructed.

A BROADER DISCUSSION:
ZONING CLASSIFICATION FOR THE RIVERFRONT

Not long after city council approved the request by Dominion Resources to build a 160-foot high office building along the river west of Lee Bridge, discussion began regarding a broader riverfront development policy and riverfront zoning. The north side of the river on the east end in the tidal reach was ripe for development, with at least two potential projects in the wings.

The discussion about how to treat this portion of the riverfront was not prompted by specific rezoning requests but kicked off by a proposal from Councilwoman Gwen C. Hedgepeth to create an RF zoning classification limiting new buildings to 150 feet in height.[214] The targeted area for RF zoning was the north-side shoreline from 18th Street to the city's eastern boundary with Henrico County, although the specific area to which the zoning classification would apply was not specified. City council, in a 5–4 vote, gave preliminary approval for creating an RF zoning classification in a resolution on April 22, 2002. The resolution directed the planning commission to hold a "public hearing on the proposed ordinance for the city's riverfront area on the north side of the James River from 18th Street east to the city's corporate boundaries, and to submit its recommendations and any explanatory materials to the City Council."[215]

The planning commission took up city council's request for recommendation with an aggressive approach. It recommended against approving the RF zoning district, which would allow building to 150 feet with few river setbacks or other restrictions. Instead, the commission proposed a package of three ordinances as an alternative. The objection to RF zoning frequently expressed during hearings was the likelihood that it would result in a "concrete canyon" or, phrased differently, a "150-foot wall along the length of the city's riverfront" that would block views from Church Hill and elsewhere.[216] RF-1 and RF-2 zoning proposed by the planning staff and the commission included requirements that would, to a greater extent than RF, protect the viewshed and provide pedestrian access along the river's shoreline. The third element of the package, conditional zoning, would add flexibility to the zoning process.

The planning commission endorsed the three-ordinance package at its meeting on June 17, 2002. City council adopted these ordinances a week later.[217]

Three key provisions of the approved ordinances made them acceptable to those opposed to the defeated RF proposal. One, of course, was height. While RF allowed 150-foot structures along the river, RF-1 restricted those closest to the river to 60 feet and RF-2 enabled buildings farther back where they would not block as many views to rise to 150 feet. Second, RF-1 and RF-2 required a setback from the river's edge ("no building or structure shall be located within 50 feet of the mean high water level") allowing space for walkways and other riverside activities. Third, the approved ordinances required "view corridors"—space between buildings preventing a continuous view-blocking wall.

Since the three approved ordinances are abstract "planning tools" that do not, without further action, result in permission to develop, it might seem surprising that commission and council proceedings stirred turnouts as large and as passionate as they did. The answer, in part, may be the seeming imminence of new developments along the river. These prospective developments, as well as the recently concluded fight over Dominion Resources expansion, no doubt gave the consideration of riverfront "planning tools" added salience. The view from Libby Hill Park provided added incentive to get the riverfront development rules right.

THE VIEW THAT NAMED RICHMOND

In a ceremony on March 18, 2006, the mayor of Richmond upon Thames, Cllr. Robin Jowit, and the mayor of Richmond, Virginia, L. Douglas Wilder, dedicated a plaque designating "The View that Named the City." The plaque reads:

> *The curve of the James River and steep slope on this side are very much like the features of the River Thames in England, at a royal village West of London called Richmond upon Thames.*
>
> *William Byrd II, an important planter, merchant, politician and writer, was asked by the House of Burgesses to plan a town at the Falls of the James in the early 1730's.*
>
> *As he had traveled several times to Richmond upon Thames, it is believed that the view led him to name this new town "Richmond."*

Both before and since the dedication of that plaque, there has been a concerted effort to protect this view of the James. Scenic Virginia, a statewide nonprofit organization, has been a leading advocate in many state and local forums for preserving and enhancing this iconic view. Several months after

179

the two mayors dedicated the plaque, Scenic Virginia, with the leadership of executive director Leighton Powell, selected the view of the James River from Libby Hill Park as its 2006 Endangered Viewshed. "This was an easy choice," says Scenic Virginia president Eugenia Anderson-Ellis. "The view from Libby Hill Park has scenic, historic, and cultural significance that extends beyond our capital city. It belongs to all Virginians, and we should take every precaution to preserve it."[218]

The primary threat is potential high-rise development that would disrupt or block the view. The view at Richmond upon Thames in England was protected by the Richmond, Petersham and Ham Open Spaces 1902 Act of Parliament; its sister view in Virginia has never had similar protection. The city, with this portion of the riverfront in mind, had established RF-1 and RF-2 zoning in 2002 (for which Scenic Virginia had given the city its Best Preservation of a Scenic River Corridor Award in 2003), but this was far from the blanket protection afforded to Richmond upon Thames. Continuing vigilance and action would be required in the Richmond, Virginia case.

As part of its campaign to protect what it began to call "The View," Scenic Virginia solicited the support of the Virginia General Assembly. Any full prohibition of development in the viewshed area, following the British Parliamentary example, would have been a nonstarter, so a resolution celebrating the Libby Hill view was proposed. The nominal occasion for the resolution was the upcoming visit to Virginia by Her Majesty Queen Elizabeth II. House Joint Resolution No. 658 was passed on February 2, 2007.

Although General Assembly resolutions by their nature have no enforcement power, some of the language in early drafts became the subject of objections by developers. To facilitate passage, several elements of the initial draft were removed and a new item added. Removed were these:

WHEREAS, the City of Richmond is currently undertaking efforts to preserve the historic panorama view of the James River from Libby Hill Park

and

WHEREAS, being of such vital importance to the history of Virginia and its capital city, the panoramic view of the James River from Libby Hill Park belongs to all of the citizens of the Commonwealth

and also

180

…which should be preserved, unencumbered and unobstructed for all of the future generations of the Commonwealth to enjoy.

Added in the draft that passed was this element:

RESOLVED FURTHER, That this resolution shall not affect local land use approvals with respect to this view…

Of course, from the perspective of Scenic Virginia and other view preservation advocates, the added phrase was the exact opposite of the hoped-for impact of the resolution. The hope was that the resolution would help prevent land use approvals that would block the view they wished to protect. Nevertheless, it was thought to be the price required to achieve passage, and Scenic Virginia hoped that the weight of the remainder of the text would overwhelm this element.

The resolution in full reads:

COMMONWEALTH OF VIRGINIA
GENERAL ASSEMBLY

HOUSE JOINT RESOLUTION NO. 658

Celebrating the panoramic view of the James River from Libby Hill Park in Richmond, Virginia.

Agreed to by the House of Delegates, January 30, 2007
Agreed to by the Senate, February 21, 2007

WHEREAS, William Byrd, the founder of the City of Richmond, was familiar with Richmond-upon-Thames in England and the view of the River Thames from Richmond Hill; and

WHEREAS, William Byrd witnessed the panoramic view of the James River and its remarkable similarity to the viewshed in England's Richmond-upon-Thames; and

WHEREAS, historians believe that in 1733 William Byrd named Richmond because of this similarity of the scenic panorama along the James River to the view in Richmond-upon-Thames; and

WHEREAS, the scenic view in Richmond-upon-Thames is recognized as a great landscape icon of England and was the first and only view to be safeguarded by an Act of Parliament, the Richmond, Petersham and Ham Open Spaces Act, 1902; and

WHEREAS, a historic plaque located at Richmond-upon-Thames in England refers to the beautiful viewshed's role in the naming of Richmond, Virginia; and

WHEREAS, the panoramic view of the James River from Libby Hill Park is recognized as a great landscape icon in both the Commonwealth and in Richmond, its capital city, and was designated by a plaque as "The View That Named the City" by the mayors of Richmond and Richmond-upon-Thames on March 18, 2006; and

WHEREAS, the panoramic view of the James River from Libby Hill Park in Richmond possesses scenic, historic, and cultural significance that extends far beyond the capital city, attracting visitors from around the world; now, therefore, be it

RESOLVED by the House of Delegates, the Senate concurring, That the General Assembly celebrate the panoramic view of the James River from Libby Hill Park in Richmond, Virginia for its historic and critical association with the early development of the City of Richmond and its connection to Richmond-upon-Thames; and, be it

RESOLVED FURTHER, That this resolution shall not affect local land use approvals with respect to this view; and, be it

RESOLVED FINALLY, That the Clerk of the House of Delegates prepare a copy of this resolution for presentation to Queen Elizabeth II, who will visit the Commonwealth of Virginia in May 2007 in conjunction with the 400th anniversary of the founding of Jamestown.

House Patrons: McClellan, Armstrong, Bulova, Eisenberg, Englin, Hall D. Jones, Lewis, Marsden, McEachin, O'Bannon, Peace, O. Ware

Senate Patron: Marsh

An objection that developers sometimes raise in the debate over land use that would obstruct The View is the argument that William Byrd II never actually proclaimed that the Richmond-upon-Thames view was his inspiration for naming Richmond. The story is simply a myth, not fact, they say, so the Libby Hill view does not deserve the attention and deference it receives.

This myth or fact question was examined as part of the viewshed study by the Virginia Chapter of the American Society of Landscape Architects. The study found nothing that documented Byrd's sentiment regarding the naming of Richmond. In 1796, Benjamin Latrobe provided the first available written comparison of the views with the suggestion that Richmond's name was the result of that similarity.[219] Over the decades, writers have concluded the same. So, whether myth or fact, The View has created a link between Richmond upon Thames, England, and Richmond, Virginia, a link that is celebrated with recurring visits and ceremonies on both sides of the Atlantic.[220]

The river is a continuing inspiration for photographic artists. *"James River Sunset" by Harold Lanna, courtesy of Scenic Virginia.*

Bridge reflections. *Courtesy of Scott Adams.*

Rocks in the river as art. *Courtesy of Scott Weaver.*

The maintenance and enhancement of the view from Libby Hill Park (and from many locations in Church Hill) depends on the uses made of private property between the park and the river. One major obstruction was removed when the city purchased the Lehigh Cement property and removed tall silos. That property is now part of riverfront park plans. Several development proposals have been made for other privately owned sites, but none formally approved or begun. USP/Echo Harbour, a five-acre riverfront parcel adjacent to Great Ship Lock Park, is pivotal to the protection of this view. Parcel owners have proposed a high-rise structure; view advocates hope that it will become a public park (see chapter 9). The outcome of this struggle has yet to play out.

While panoramic viewsheds (including some of those portrayed in the color insert) deservedly receive the most attention for protection, immediate, less extended views add to the enjoyment the river offers. In many respects, these closer views provide greater opportunities for "photographic artistry," capturing a special scene at a unique moment.

PERSPECTIVE

Scenic vistas abound along the James River in Richmond, but it is unlikely that the Virginia General Assembly or the Richmond City Council will provide the kind of viewshed protection that the long-standing Act of Parliament offers to Richmond-upon-Thames in England. Yet the three cases described indicate that protecting viewsheds has been on the radar of advocacy groups, planners and public officials. That vigilance continues. Scenic Virginia, for example, is developing a program to help citizens identify their favorite views and vistas and convey that information to public decision makers,[221] and the 2012 Riverfront Plan gives prominent attention to view protection.

The establishment of James River Park System and other public places provides public access to viewpoints and serves to protect the views they encompass, and the James River Park Conservation Easement reinforces the protection the park provides. Still, the public places do not line the entire riverfront or the property between key viewpoints and the river; development could block valued public views. Although the private sector can be an enemy of public vistas, it can also enhance access to those views. In many waterfront cities around the world, restaurants were among the first

Not long ago, no restaurants in Richmond provided a river view—now some do, with more to come. This one is from Rocketts Landing. *Photo by author.*

to offer a public view of water vistas. For decades, there were no such public restaurants along the James River in Richmond. Now several restaurants do offer a river view, and there are other opportunities for enterprising restaurateurs who are able to navigate the web of property ownership.

While the future seems bright for protecting the Richmond river views, scenic advocates are not likely to get all that they might want. The Riverfront Plan suggests some limits to maximizing public views in its discussion of "balance" between public and private control. Since "balance" also means "satisfying conflicting interests," it may be a euphemism for battles yet to come. The cases outlined here demonstrate that viewsheds and access to them are not guaranteed; vigilance and struggle may be required to ensure their protection. Scenic Virginia, Falls of the James Scenic River Advisory Committee, the Virginia Chapter of the American Society of Architects and others vested in protecting viewsheds will continue to have work to do.

The enjoyment of seeing beautiful and interesting views was one of the earliest forms of recreation along the river and still is the most easily accessible recreational use of the river. View gazing does not require the paddling skill or the physical exertion or the expenditure of time that other forms of recreation demand. Of course, some of the best, most inspiring and sometimes surprising views are encountered while paddling, hiking or biking. Whether sedentary or active, fleeting or extended, river views provide pleasure and sometimes even inspiration.

The growing display of history along the riverfront, like the many engaging river views, provides a readily accessible recreational and educational opportunity.

SHOWCASE FOR HISTORY

RIVERFRONT AS OUTDOOR MUSEUM

Richmond's early history was also national history.
—Marie Tyler-McGraw, At the Falls

*The drama that has played itself out along the shores of the James
is as powerful a tale as has ever been told.*
—Bob Deans, The River Where America Began

On July 30, 2007, U.S. House Resolution 16, sponsored by Representative Jo Ann Davis, recognized the James River as "America's Founding River," a designation that is indicative of the history that has played out along its length. The James River Association continues to highlight that designation. Much of America's history from the first English settlement at Jamestown through the Civil War was James-centric. At least two books have used the James River as the organizing spine for recounting the history of the United States. The earlier of the two is *The James* by Blair Niles, later revised and expanded as *The James: From Iron Gate to the Sea*. The more recent is Bob Deans, *The River Where America Began: A Journey Along the James*. The current riverfront along the James in Richmond displays much of that history.

At almost any point along Richmond's ten miles of the James, a visitor will encounter a historical marker, memorial, artifact or architectural remnant. Some are major historical structures like the Pump House, or Tredegar Iron Works, or Tidewater Connection Locks, or Great Ship Lock. Others are simply signs or medallions in the walkway describing historical scenes or events, like the Haxall Millrace or the Belle Isle Civil War Prison. Still others are monuments like the Christopher Newport Cross or the statue of Abraham Lincoln and his son Tad.

Some history lovers and preservationists have worked hard, sometimes successfully and sometimes not, to preserve structures like the Washington Arch, James River and Kanawha Canal (see chapter 7), Tredegar Iron Works and Byrd Park Pump House. In addition to preservation, special efforts have been made to include historical interpretation in riverfront developments. On the Canal Walk and elsewhere, the variety of monuments, historic medallions and interpretive signs that have been erected along with the historic structures that remain have produced an outdoor historic museum along the river that continues to grow. The James River in Richmond, indeed, has become a showcase for history.

A full recounting of the history along the river just in Richmond would require a book in itself—or several. This far less ambitious chapter simply is intended to offer an introduction to the outdoor historical museum that is being preserved and created, along with a few stories about the conflict, cooperation and passion that have accompanied that process. Let's sketch what a visitor might find traveling upstream along the banks of the river.

A RIVERBANK SKETCH OF HISTORY—NORTH SIDE

Orleans Street, the city boundary at the edge of the newly developed Rocketts Landing (a multiuse development in Henrico County with condominiums, restaurants and businesses) begins what can be an extensive tour of history along the banks of the river. The mile or so upstream was a bustling seaport for much of its history, the terminus for oceangoing vessels that covered the 125 miles from the Atlantic. Before the Civil War, Rocketts was a major port and trading post; during the Civil War it served as the CONFEDERATE NAVY YARD (on both sides of the river). As ships became larger, the river's depth was no longer adequate for oceangoing vessels. Lehigh Cement operated here until recently—before the city acquired the property and removed the tall silos from the riverbank in 2015. Virginia Capital Trail provides a pedestrian/cycling path parallel to the river through this area that is being redesigned as a product of the 2012 Riverfront Plan. Intermediate Terminal Warehouse No. 3 is scheduled for renovation as a restaurant

A view of the Confederate Shipyard at Rocketts, circa 1864. *Courtesy of the Library of Congress.*

and beer garden. The destiny of the privately owned riverfront property just upstream of the terminal and the former Lehigh Cement site is still undetermined—the owners are considering building here, while others are hoping that it becomes a public park.

At the foot of Pear Street, GREAT SHIP LOCK is a significant site in the river's outdoor history museum. Completed in 1854, although preceded by several less durable locks dating back to 1816, the Great Ship Lock was built to lift oceangoing vessels from the river into RICHMOND DOCK, where cargo was transferred to canalboats and wagons.[222] The dock was part of a link with the Great Turning Basin and the Kanawha Canal and now connects to the restored canal through a gate in the floodwall. (The historic name "dock" is sometimes confused since it now appears simply to be a large canal and does not match the usual image of a dock.) It is the dream of some to return the lock to working order and serve as a connection between the river and the Richmond Dock for large pleasure craft. One big obstacle, in addition to funding, is the Norfolk-Southern bascule bridge that crosses the channel quite close to the lock. The bridge is still used but, without major repair, is no longer capable of being raised.

Visible from Great Ship Lock over the rail trestle and the trees at the top of the bluff is the CONFEDERATE SOLDIERS AND SAILORS MONUMENT located in Libby Hill Park (the best site from which to see the "view that named a city," see chapter 10). Here stands a seventeen-foot soldier atop a ninety-foot granite column. The statue was unveiled on May 30, 1894, with a reported audience of 100,000.[223]

From the Great Ship Lock, the Virginia Capital Trail extends along the Richmond Dock to 17th Street and merges with the Canal Walk. This walking/cycling path is primarily under the elevated CSX (formerly C&O) rail line, itself a historical structure. This railway viaduct was built in 1901 to traverse a difficult two miles along the route through Richmond[224] and is credited with being the longest elevated double freight track in the world.

The historical display intensifies at the "spaghetti works," an area between 17th and 14th Streets where the Canal Walk, the floodwall and Capital Trail come together and where the canal passes through the (fought for) floodwall gate. The aptly, but unofficially, labeled spaghetti works is a combination of highway overpasses, rail lines, Canal Walk, floodwall, parking lot and retail establishments. Here historical markers abound, although the river itself is not visible since it is behind the floodwall. This is the location of the renowned "TRIPLE CROSSING" where three rail lines intersect, one atop the other. From this point west on the Canal Walk, a visitor is presented with a continuing and varying array of historical displays and orientation pylons. The Canal Walk is sprinkled with twenty-three bronze medallions inlaid in the walkway with historical information.

Among the simplest, yet most dramatic displays are the plaques on the floodwall depicting the heights of the four LARGEST FLOODS that ravaged Richmond. The highest is Historical 1771, followed by Agnes 1972, Juan 1985 and Camille 1969 (see chapter 8). The heights the water reached are indeed striking, making the millions of dollars of damage these floods caused seem quite plausible.

A display, not for the claustrophobic, is the replica of the crate used by "BOX BROWN" with the description: "In a wooden crate similar to this one, Henry Brown, a Richmond tobacco worker, made the journey from slavery to freedom in 1849." He made the twenty-seven-hour trip from Richmond to Philadelphia with crackers and a pouch of water—and the "This Side Up With Care" instructions were not always followed.[225]

WASHINGTON'S VISION is a large display of the James River and Kanawha Canal from Richmond to Buchanan describing the canal that so significantly contributed to Richmond's growth as an industrial and commercial center in the first half of the nineteenth century. The NEW TURNING BASIN (not in the location of the historic Great Turning Basin, which now is covered

A replica on Canal Walk of the container that carried Henry "Box" Brown from Richmond to Baltimore in 1849. *Photo by author.*

The Great Turning Basin in downtown Richmond, 1865. *Courtesy of the Library of Congress.*

by high-rise buildings) provides a place for tourists to catch the boats for a historical narrated cruise along the eastern portion of the restored canal and Richmond Dock.

The use of the riverfront as an outdoor museum has not always been without controversy. An extended and passionate public debate arose when vinyl mesh murals depicting Richmond history were displayed on the floodwall adjoining the Canal Walk. Sponsored by the Richmond Historic Riverfront Foundation, the display included twenty-nine images around themes of war, floods, power, transportation and more. One of the images was that of Robert E. Lee in Confederate dress uniform. City councilman Sa'ad El-Amin objected passionately and very publicly to the inclusion of the Lee mural. After a good deal of public debate, including a threatened boycott, a new combination of images was approved by a special committee. In this new set of murals, the Lee mural was replaced with his image in civilian rather than military dress. Not long after it was hung, the new image was burned by vandals but then replaced again.[226]

Another critique of the floodwall murals and the Canal Walk outdoor museum was not as heated, nor did it stir up public controversy, as the attack on the Lee mural did. In the lead article in the Commentary section of the *Richmond Times-Dispatch*, Dale Wiley argued that the Canal Walk should focus its historic display on the canals themselves, especially the James River and Kanawha Canal. What is in place, Wiley argued, is a set of "banners displaying many individuals who have little or no historical relationship to the canal, its origins, its engineering, its construction, or its use." He went on, "Richmond has an asset few cities in the country can boast: a truly historical waterway that was started in Richmond. Such a gem of history should not be wasted by using it for other reasons."[227]

The ultimate fate of the floodwall mural project as an outdoor museum did not blossom or die in the Lee controversy but rather faded away (quite literally) before it was removed. Nor did Wiley's proposal to focus exclusively on the canal win the day—other topics on the Canal Walk as museum continued to be displayed. Nevertheless, the Kanawha Canal and Haxall Canal histories do now make up a large part of the Canal Walk exhibition. A portion of the floodwall is scheduled to be the "canvas" for a street art mural festival in 2020,[228] so new color will be added to the wall, with or without controversy.

The Canal Walk has two elevations connected by a set of stairs. A walk up the stairs provides a dramatic indication of the height the historic locks lifted or lowered canalboats. Unfortunately, the development of the new canal did not establish a navigable connection between the elevations, although two of the historic locks are preserved. It would be expensive, but such a connection is still feasible.

The two preserved TIDEWATER CONNECTION LOCKS beneath Reynold's Wrap Distribution Center offer a glimpse of Richmond and Virginia history. Several descriptive signs, including this one, offer interpretation:

> *In 1854 the Tidewater Connection Locks linked the canal basin to the James River tidewater below Richmond. The system contained five granite locks, each measuring 15 x 100 feet. This resulted in a flight of water stairs that lowered boats a total of 69 feet within a distance of 3½ blocks.*

After one climbs the stairs to the upper canal level on the way to Brown's Island, a monument depicting much earlier history is found. The CHRISTOPHER NEWPORT "CROSS AT THE FALLS" now stands near the beginning of the upper Canal Walk. The historic event commemorated is described by Virginius Dabney in the opening paragraph of his history of Richmond: "One-armed Captain Christopher Newport had led a small band of intrepid English explorers upriver from Jamestown—following their epoch-making landing there ten days before—and planted a wooden cross at The Falls, near the heart of today's downtown Richmond. It was May 24, 1607."[229] Claiming the area for King James, Newport is thought to have placed the cross somewhere near the current location of Mayo's Bridge. The commemorative cross was originally placed on Gambles Hill and moved twice, finally to its current location on the Canal Walk near 12th Street and Byrd Avenue.[230]

Entering Brown's Island from the east, the Canal Walk passes right alongside the restored Haxall Canal (not the Kanawha Canal) through the shell of the old hydroelectric plant, now decorated with "futuristic murals." The remnant of the plant is a reminder of the little-known fact that Richmond was the home of the first commercially successful ELECTRIC TROLLEY in the United States, which operated from 1888 to 1949.

The Canal Walk on BROWN'S ISLAND, like the section at spaghetti works, is dotted with medallions and displays. As one of those medallions informs, the partially man-made Brown's Island "was created when the Haxall Canal was extended west to the Tredegar Iron Works." Now owned by the city and managed by Venture Richmond, Brown's Island was the center of Richmond's industry for two centuries. The HAXALL MILLRACE is one example of that industry.

The first gristmill in Richmond was built on rocks in the river and approached by planks laid from one rock to another.

In the 19ᵗʰ century fleets of schooners and brigs carried Richmond flour to Brazil and around Cape Horn to San Francisco and Australia.

From colonial times the waterpower of the James was used to grind wheat into flour. This became even more effective when millraces, like the Haxall Canal, were dug to divert water from the river for this purpose. Eventually, Richmond became one of the largest flour exporters in the world.

Other medallions describe MANCHESTER & FREE BRIDGES, JOHN JASPER, INDIAN HISTORY ALONG THE RIVER AS FAR BACK AS 10000 BC, TREDEGAR IRON WORKS, and many others.

Nearing the upstream end of the Canal Walk is *The Headman*, a statue by Paul DiPasquale depicting a black bateau operator in action at the oar. This statue commemorates the contributions of African American men as skilled boatmen on the James River and its canals and in the development of industry and commerce in the City of Richmond.

The culminating exhibit at the west end of the Canal Walk at the gateway to the T. Tyler Potterfield Memorial Bridge (formerly Brown's Island Dam) is the April 1865 Exhibit, a series of quotations from the evacuation of RICHMOND IN THE LAST DAYS OF THE CIVIL WAR. The introductory signage reads:

ALONG THIS BRIDGE, the events of the first week in April—when Richmond, the capital of the Confederacy, fell to the Union army—are recounted in the words of people who were present at the time.

Just upstream from the western entrance to the Canal Walk, a visitor has the opportunity to immerse in Civil War history as well as witness an important aspect of nineteenth-century industrial activity along the river. The remnants of Tredegar Iron Works now house two Civil War museums: the AMERICAN CIVIL WAR MUSEUM AT HISTORIC TREDEGAR and the RICHMOND NATIONAL BATTLEFIELD PARK VISITOR CENTER.

Particularly in the 1800s, this section of the river was a bustling manufacturing center, relying largely on water power from the canals. For well over a century, TREDEGAR IRON WORKS, founded in 1836, manufactured an array of products, including locomotives, train wheels, cannons and

April 1865 Exhibit at the entrance to the T. Tyler Potterfield Memorial Bridge. This exhibit predated the construction of the bridge. *Photo by author.*

ordnance, armor plating for ships and much more. Tredegar was the chief source of manufactured iron products for the Confederacy.

Behind the Tredegar buildings is a life-size sculpture of ABRAHAM LINCOLN AND HIS SON TAD in front of the inscription *To Bind Up the Nation's Wounds.* This display, another that caused some controversy when it was placed, commemorates Lincoln's visit to Richmond just after its fall in April 1865.

Not far from Tredegar, a stroll across the cable-supported pedestrian walkway beneath the Lee Bridge provides great views of the river and leads to Belle Isle, another "room" in this outdoor museum, one that shares space with joggers, mountain bikers, rock climbers, sunbathers and paddlers. "From a Powhatan Indian fishing village to a 20th-century steel plant, this 65-acre island in the heart of Richmond shows the social and economic history of the city in microcosm" says *An Interpretive Guide to Belle Isle.*[231] And so it does. Belle Isle history ranges from prominent owners and renters (William Byrd I, II and III, Benjamin Latrobe, Bushrod Washington, Lighthorse Harry Lee), to its use of water power, then steam,

Tredegar Iron Works in April 1865 at the conclusion of the Civil War. *Courtesy of the Library of Congress.*

then electricity for industry, to its role in national history as an infamous CIVIL WAR PRISON CAMP, to its current incarnation as a central feature of the James River Park System.

Even though the unrestored remnants of OLD DOMINION IRON & NAIL WORKS (another important supplier for the Confederacy) are still visible, as are a former GRANITE QUARRY and the remnants of an early twentieth-century HYDROELECTRIC PLANT, the history of Belle Isle is much richer than the existing artifacts and interpretive signage might indicate. Most of the industrial structures have been removed and no physical features remain of the prison camp that housed enlisted Union prisoners. Still, the physical setting, with a little mystery and some unanswered questions, prompts the imagination to envision what occurred here. Perhaps a little mystery is appropriate—even the source for the name "Belle Isle" is uncertain.

One story is that the island became known as Belle Isle when the Belle Isle Manufacturing Company (predecessor to Old Dominion Iron & Nail Works) was chartered by the Virginia Assembly in 1832. Another version, a bit more entertaining, is that the island was named for James Bell, who reportedly ran a rather disreputable horse racetrack on the island. The genteel ladies of Richmond, in an effort to retrieve the island's reputation, however, changed the spelling from Bell to the more refined "Belle."[232] And clearly *isle* has more gentility than *island*.

Visible across the river from the northwest side of Belle Isle is HOLLYWOOD CEMETERY, a part of Richmond history since its dedication in 1849. Its founding provoked a bit of controversy. "It was claimed that the promoters were merely trying to make money and that corpses buried in proximity to the city water works would contaminate the municipal water supply. Some also contended the 'noise and tumult of the falls' would disturb the dead."[233]

Hollywood Cemetery provides a double payoff for visitors: a rich dose of history and dramatic views of the river from several locations. The cemetery contains the remains of two U.S. presidents, James Monroe and John Tyler, as well as Confederate president Jefferson Davis. (Raft guides in their dry— or is it wry—humor like to say that two and a half presidents are buried here.) Also here are the remains of hundreds of Civil War soldiers and officers. Some were casualties of war; others were survivors who were buried in Hollywood after the war was over. The remains of many other dignitaries, as well as ordinary mortals, are here.

The view from Hollywood Cemetery has been a continuing point of praise. Mary H. Mitchell describes the view as it was when the site was selected for the cemetery:

> *In the distance…one could see the steeples of Richmond's many downtown churches, the dome of city hall, and the classical lines of the State Capitol building. South of the proposed cemetery site, at the base of the bluff, lay the James River and Kanawha Canal. A wide towpath followed the contour of the canal's southern bank; from the towpath the shore sloped gently to the river. Two railroad bridges and Mayo's toll bridge crossed the James east of Harvie's woods, but no man-made obstructions spoiled the southern vista; the falls of the majestic river and verdant Belle Isle.*[234]

Today there are more man-made obstructions, so church steeples and the capitol are blocked by office buildings. But the view is still dramatic.

Not far upstream from Hollywood Cemetery a visitor finds another site of American and Richmond history. BYRD PARK PUMP HOUSE PARK provides a cluster of historical remains begging for preservation. Three parallel canals represent three historic periods. Closest of the three to the river is a remnant of the JAMES RIVER CANAL, the first canal system in the United States, organized in 1785 and opened in 1789 while George Washington was still honorary president of the canal company. The most striking feature of the canal's remains is the Lower Arch, sometimes called WASHINGTON'S ARCH. According to Bill Trout, Washington visited the site during construction and

A current view from Hollywood Cemetery. *Courtesy of Scott Adams.*

rode through in a bateau in 1791.[235] The James River Canal allowed bateaux to bypass the rapids of the fall line, but the remainder of trips up- and down-river were in the main river channel.

Closely paralleling the short remains of this canal is the JAMES RIVER AND KANAWHA CANAL, which was built beginning in the 1820s to extend canal transportation westward. Two of the many locks of the Kanawha Canal are at this location and are known as THREE MILE LOCKS since they are three miles from the Great Turning Basin downtown. According to Trout and others, these locks are capable of restoration to operating condition.

Just north of the Kanawha and parallel to it is the PUMP HOUSE CANAL, built to run water to a twenty-foot drop that powered pumps to push Richmond's drinking water to the reservoir in Byrd Park. BYRD PARK PUMP HOUSE is a striking Victorian Gothic–style structure of granite and cast iron built between 1881 and 1883. The impressive structure seems entirely out of its natural habitat, resembling a courthouse or a church more than a pump house. The upper floor is an open pavilion remembered for dances and social events held into the 1920s.

The grounds are replete with interpretive signage. Here is a portion of one historical note:

The large granite building is what remains of the second oldest water pumping station in Richmond.

It was built in 1882 and provided drinking water for the City of Richmond for over forty years. Nine pumps on the first floor drew water from the canal in front of you and sent it up the hill behind you to the Byrd Park Reservoir at Blanton Avenue and the Boulevard.

The hope of canal historians and preservationists like Bill Trout, Jimmy Moore, Lynn Lanier and others has been to restore the locks to working order and convert the area to a major Virginia canal museum. The history and grandeur of the Pump House have stirred other individuals and groups to seek new uses—museum, retreat center, restaurant, meeting hall. Although a new roof has been installed to protect the Pump House from further damage and a great deal of restoration has been accomplished by the James River Park and a multitude of volunteers, the grand aspirations have not yet been achieved. Perhaps this will change. Friends of the Pump House has taken on new energy and the new park master plan gives the Pump House a prominent place. Even if these grander hopes for the Pump House and adjacent canals are not fulfilled, the surviving artifacts along with the trails, bridges and interpretive signage make this spot a signature part of the river's outdoor museum, a place well worth visiting.

Significant historical sites, along with the city's water treatment system, lie along the north bank of the river from Pump House Park upstream to Bosher's Dam, but they are largely inaccessible. The most prominent is the JAMES RIVER AND KANAWHA CANAL, still watered and used as backup for the city water supply. Much of its length is navigable for small craft but not publicly accessible. Some still hope that this reach of the canal, as well as that from Tredegar to Pump House, can be made available for recreation as well as historical display: "Someday, we need a safe canoe portage and a lock for batteaux and for tour boats going to Bosher's Dam."[236]

The remains of the VILLAGE OF WESTHAM and the associated iron foundries are unexploited archaeological sites but on private property and accessible only by permission. Westham was chartered by the Virginia General Assembly in the mid-eighteenth century as a shipping and trading post to receive tobacco and other products shipped downstream from farms to the west. To avoid the rapids which begin at this point, loads were taken from bateaux and hauled over dirt roads to Richmond. In addition to transport, manufacturing was nearby: "The largest manufacturing operation in the Richmond area during the Revolution was Westham foundry. It was established and owned by the State of Virginia and located on the "north bank of the James River about one mile below the village of Westham and slightly downriver from Williams Island."[237]

Near the Westham iron foundry, WESTHAM ARSENAL achieved enough importance to be the focus of a raid in the Revolutionary War ordered by Benedict Arnold and carried out by Lieutenant Colonel John Graves Simcoe. Thomas Jefferson personally organized the evacuation of the arsenal, and he and the troops left just hours before Simcoe arrived. A later foundry, Westham Iron Works, supplied iron to Tredegar Iron Works during the Civil War.[238]

A RIVERBANK SKETCH OF HISTORY CONTINUED— SOUTH SIDE

Returning to the starting point at Orleans Street, but on the south side of the river, historical sites continue. By choice or accident of history, the city of Richmond was located on the north side of the river, so fewer events of historical note occurred on the south side. Although this bank of the river is not as densely populated with historical structures and interpretive signage, it has been the site of some of Richmond's most poignant history, both recent and more distant.

Across the river from Rocketts Landing is Manchester Dock. The large concrete boat ramp there is a reminder of Newton Ancarrow, one of the earliest and strongest advocates for cleaning up the James River in Richmond. A man some call the "city's first environmentalist," he was an early pioneer in the transformation of the river.[239] The location is ANCARROW'S LANDING, the only launch site in the city for motorized craft. Ancarrow was a builder of speedboats sold around the world; among his clients were the emirs of Kuwait and Bahrain, King Paul of Greece and Aristotle Onassis. He also was an irritant to city officials. "Ancarrow testified about the river's health to an apathetic city council around 1966. He brought a large jar filled with putrid water, in which floated a condom and a dead rat. Council dismissed the evidence."[240] (As mentioned earlier, this story varies in the telling.) Nor was he hesitant to take matters to court. He sued the city, the EPA and the Army Corps of Engineers, among others. And usually lost.[241]

The city, with the prodding and help of the federal and state governments, did eventually clean up the river although not at a speed that

The boat launch at Ancarrow's Landing, the only launch site for motorized craft within the city limits. *Photo by author.*

suited Ancarrow. And the city, some say exhibiting "meanspiritedness," condemned Ancarrow's property to become the location for the wastewater treatment plant. The landing area itself is now part of the James River Park System and the starting point for a memorial of a dark side of Richmond's history.

Just upstream of Ancarrow's Landing at Manchester Dock is the beginning (or end) of the RICHMOND SLAVE TRAIL, a series of historical markers that follow the path of slaves from ship to the slave markets and auction houses on the north side of the river (or in reverse at a later time). "Designed as a walking path, the Richmond Slave Trail chronicles the history of the trade in enslaved Africans from their homeland to Virginia until 1778, and away from Virginia, especially Richmond, to other locations in the Americas until 1865."[242] The trail was developed by the Richmond Slave Trail Commission, established by the Richmond City Council in 1998 to help preserve the history of slavery in Richmond.

Richmond was one of many destinations in the Americas for slaves shipped from Africa until Virginia prohibited further importation in 1778. As economic conditions changed, especially with the increased demand for labor in the cotton fields of the Deep South, Richmond transitioned from an importer to an exporter of slaves. Manchester Dock was a primary location for loading slaves onto ships for the trip to New Orleans and other destinations. Apparently, the phrase "sold down the river" originated here.[243] By mid-eighteenth century, the role of the river declined as the major form of transport of "surplus" slaves. Still, the marketing of slaves continued until the Civil War, and by one account, the domestic slave trade was the single largest piece of Richmond's economy for the period from the Revolutionary War to the Civil War.[244]

MANCHESTER, an independent city before it merged with Richmond in 1910, was a major manufacturing site as well as a transportation hub connecting rail and river. Tobacco, coal from the Midlothian mines and the aforementioned slaves were shipped through Manchester.

For a period of time, the river and the canal were intense competitors with railroads for the transportation business. Rail, of course, won. The close connection between river and rail makes it appropriate that the RICHMOND RAILROAD MUSEUM sits in close proximity to the James River, just at the south end of Mayo's Bridge.

MAYO'S BRIDGE (aka 14th Street Bridge) is a noteworthy piece of Richmond history. Prior to the first toll bridge in 1788, Coutts' Ferry was the only connection between the growing city of Richmond and the town of Manchester on the south side. The bridge had a rough existence; it was destroyed multiple times by ice flows and floods and then burned by

Mayo's Bridge today. Various iterations date back to 1788. *Courtesy of Scott Adams.*

Richmond and Petersburg Railroad Bridge piers. The taller, older, more enduring piers were built in 1838. *Photo by author.*

retreating Confederates in the evacuation of Richmond in 1865. In 1910, Richmond purchased Mayo's Bridge, until then constructed from wood, and entirely rebuilt it with concrete. Since its reopening in 1913, it has withstood multiple floods, including the Agnes flood in 1972 in which the bridge was entirely covered by the floodwater.[245]

The MANCHESTER CANAL is still visible on both sides of 14th Street, although it no longer serves its original purpose of supplying water power for manufacturing in the town of Manchester.

Not far upstream, and accessible by the floodwall walk as well as a path near SunTrust, is the abutment of the RICHMOND AND PETERSBURG RAILROAD BRIDGE built in 1838. Piers in the river are still standing. The bridge burned in 1882, reopened in 1883 and was rebuilt in 1902 on smaller piers. The bridge was removed in 1970. Both sets of piers are still visible. Ironically, the older ones are still standing strong while many of the newer piers have toppled over.[246]

A thorough review would reveal far more sites and events than portrayed here; the "ghosts" of the past are everywhere along the river. If current trends continue, many more of those ghosts will be made visible.

PERSPECTIVE

Along the river and elsewhere, Richmond has struggled with the competition between preserving the old and building anew. The canals have been especially vulnerable in this struggle. The Kanawha Canal in downtown is no more, and the Great Turning Basin as well as three of the Tidewater Connection locks have been destroyed. Yet many sites have been preserved. Richmond Dock, the connection between the Kanawha Canal and the river, almost became a sewage retention basin before an outcry turned that plan aside. Two of the five Tidewater Connection locks and several locks upriver, including Washington's Arch, are still intact and capable of being restored to working order. Great Ship Lock was brought back to working condition in the late 1980s but not maintained. It could be repaired again. There still is potential for making long stretches of the Kanawha Canal available for recreation—and historical interpretation.

Some efforts at preserving sites through adaptive reuse have been successful, and others have failed or have stalled. The Tredegar Iron Works now holds two successful historical museums and an abandoned bridge abutment is an actively used climbing wall. On the other hand, Byrd Park Pump House has been the focus of any number of proposals for reuse, none of which yet has come to full fruition. Nevertheless, the building has been preserved, and work toward its renovation is accelerating.

Of course, it is not necessary to have historical structures to document the past and educate new generations. As the multiple in-ground medallions, signs and sculptures on the Canal Walk and elsewhere amply demonstrate, newly created displays can get the job done—although not quite as well perhaps. New creations raise the question of what should and should not be memorialized and how. The controversies about what appropriately should be displayed on the floodwall show that the reinterpretation of history is still an issue, as do the recent discussions of Civil War monuments elsewhere in the city. These occasional conflicts may emerge again, but they are unlikely to forestall the trend of expanding historical display and interpretation along the James River in Richmond.

This brief and incomplete sketch of the history showcased along the river is indicative of the richness and variety of that history. It might seem ironic that the modern transformation of the river includes robust recognition of the past, but that recognition only deepens the significance of the transformation now being achieved along the river. Using the river as a recreational amenity is compatible with historical preservation, display and interpretation in a way that the former industrial use could never achieve. With the modern function of the river as amenity well established and with the strong interest in history that characterizes Richmond, it is likely that more rather than fewer historic sites and events will be memorialized in the years ahead.

The stories of transformation sketched in this book, and the many more untold, signify a new way of thinking about the river and a hopeful future for the James River in Richmond.

A NEW RIVER ETHOS AND A HOPEFUL FUTURE

People protect what they love.
—Jacques Cousteau

The best way to predict the future is to create it.
—credited to Abraham Lincoln (among others)

The transformation of the James in Richmond reflects a significant change in attitudes, behaviors and beliefs about how the river should be valued, how it should be used and how it should be protected—a new river ethos. This change in thought and behavior is as important as the change in the health of the river and preserved wilderness along its banks. The new river ethos did not in any full sense precede the physical changes that have occurred along the river; the river ethos and river restoration developed in a symbiotic relationship over time. Before changes began in earnest, only a few individuals had a sense of what the river was worth and what it could become. And even those individuals have grown in their understanding of what is desirable and the limitations on the possible. As transformation moved forward, a new way of valuing the river grew with it, and an expanded and more diverse group of river users and enthusiasts emerged.

A NEW ETHOS

This was not the first shift in ethos—attitudes, behaviors and beliefs—other shifts occurred in the centuries and decades that preceded the current era.

As Woodlief and Nelson describe in *The James River as Common Wealth*, several distinct periods characterize the river's history.

"For the Indians," Woodlief and Nelson argue, "this river was the center of their lives. They lived by its rhythms....The river could not be owned or tamed, only lived with, respected, and held in stewardship for future generations."[247]

In the early years of exploration and settlement, the English saw the river as a route to riches. But the James River, as they named it, never led to gold or a route to the South Seas. So the river became a more utilitarian means for livelihood and, for some, wealth. "All they really cared about the river was how it could serve them as they conquered and civilized the wilderness."[248]

The James River, however, was not a fully cooperative partner in transportation and manufacturing. The nonnavigable falls and unpredictable water levels made a canal necessary for transport, and floods washed out bridges, ferries, chunks of the canal and much more. "Undoubtedly Virginians felt justified in resenting a river that kept causing them so much inconvenience."[249]

One convenience the river did offer was a means for waste disposal; it was cheap and handy to pipe their waste to the river, so they did. For much of the 1800s until the late 1950s, Richmond and cities and towns upstream and down piped their waste directly to the river. The river was thought to be dirty, as indeed it was, so it was ignored. Recreational trips to "the river" meant the Rappahannock or some cleaner river.[250]

The last half of the twentieth century, as we have seen, has produced another shift in attitudes and behaviors. While it would be naïve and idealistic to suggest that we now live in rhythm with the river as the Indians are claimed to have done, that is at least the direction of change. A significant element of the modern river ethos is to allow the river to return to its pristine pre-1607 state. That, of course, is only an ideal, one that cannot be reached. Nor is it even desired in its fullest sense; the modern ethos is far more diverse and full of competing perspectives.

The modern river ethos that has developed in the last half century along the James in Richmond has three indispensable elements.

- The river is valued.
- The health, naturalness and beauty of the river are promoted and protected.
- Multiple users and perspectives are respected.

VALUED

The most fundamental element in the modern river ethos is that the river, both as it is and as it can become, is valued. Without that, little else would follow. Clearly, ignoring the river, as was the case in the middle of the twentieth century, was not valuing it. For most Richmonders, the river was out of sight, so it was out of mind. Apathy is an accurate description of attitude toward the James River at that time. While most ignored the river, it was valued by a few, both for what it was and what it might be in the future. The plan to build a scenic drive along eastern Riverside Drive in the late 1940s, the sewage safari that saw potential as well as contamination, the early downtown plans seeking to make the river the centerpiece of the city, residential building on the bluffs overlooking the river and even the teenage partying at Pony Pasture are all indications that some appreciated the river's value or at least its possibilities.

While most Richmonders ignored the river and considered the riverbanks and floodplains of little value (so therefore an appropriate location for an expressway, for example), a few had other ideas. Sparks of interest here and there began to appear, and slowly, momentum toward transforming the river began to build. Over a somewhat torturous multidecade process, the James River moved from a subject of neglect to Richmond's most celebrated feature, a transition that accelerated as more individuals and organizations came to value the river. The more individuals became involved, whether playing, working or lobbying, the more they valued it. The more they understood and valued the river, the more willing they were to be involved in its reclamation and conservation. Valuing and involvement are mutually reinforcing as can be seen in the fight to prevent the construction of the expressway on the south bank of the river, the building and support of the James River Park System, the reclamation of the downtown riverfront, and more. Simply put, the modern river ethos means valuing the river, which arises from involvement producing a commitment to stewardship.

HEALTH, NATURALNESS, BEAUTY

Simply valuing the river, however, is not enough to constitute a modern river ethos. After all, the river was valued, just in a different way, when it was used for water power and waste disposal. The modern ethos requires that the river be valued for its health, naturalness and beauty and that those values be promoted and protected. The recent history along the river displays some of these attitudes and actions.

Cleaning up the river and creating a healthy ecosystem is a cornerstone of the modern river ethos. The investment of millions of dollars by the city and the federal government in a modern wastewater collection and treatment plant was pivotal. However, local citizens like Newton Ancarrow and the pressure that he put on city council to clean up the river and the James River Association's insistence that the combined sewer overflow problem could not be ignored are integral to the development of the new value given to the river. Valuing the river for its health, naturalness and beauty also is reflected in the steps to create and maintain healthy wildlife habitat, including the effort to provide passageways upriver for anadromous fish, in the struggles to maintain scenic views of the river and more. A beautiful, natural, healthy river is essential for its new use as an amenity, and valuing it as an amenity encourages efforts to keep it healthy, natural and beautiful.

MULTIPLE USERS AND PERSPECTIVES

The James River ethos, like this reach of the river itself, contains variety—a variety of users and corresponding perspectives. The ethos must accommodate and respect this diversity and make the difficult choices and compromises sometimes required. A key question is, how divergent can these viewpoints be and still be a respected part of the new value set? Several of the stories told here display that variety and some of its inherent tension.

The sense of respect for multiple users and the dilemmas it sometimes presents is captured well in a letter written by Dr. R.B. Young to an individual who had complained about some activity along the river: "In all matters I personally have felt a deep sense of responsibility to the public (who I would like to see properly enjoy the river), to the riverfront landowners (who must bear the brunt of the abuses which occur), to the conservationists (who would like to see the area preserved in a completely natural state) and to the city officials (who must try to accomplish all of these objectives in the most practical way)."[251] That sense of respect for the multiple users is part of the river ethos that leans in the direction of "public regardingness," but it also is a respecter of private rights and values.

Respecting multiple users and different perspectives can be a difficult task. What are the limits? How tolerant of divergent perspectives can the ethos be without breaking down and becoming meaningless? Satisfactory answers may be difficult, but not impossible. The specific shape of this collaboration/accommodation will continue to evolve.

THE NEW RIVER ETHOS HAS ITS CHALLENGES

Modern river values provide support and protection for the river, but do not offer guarantees; there still are challenges.

A fundamental issue, perhaps one with increasing urgency, is whether the river experience will be ruined by overuse? At some point, the number of visitors may cause a decline in the health of the ecosystem or reduce the quality of the user experience. It is reasonable to ask whether this will be a self-regulating problem or whether policy and management actions are required to limit the volume of users.

In many respects, the river has become the centerpiece of the city that some ambitious early planners and advocates envisioned. This attention is an indication of the high value placed on the river, so one might reasonably assume that it also ensures continued protection of the river. However, there is more than one way that value can be realized. One way would be for the city to seek to maximize the revenue the river can generate and do it in a way that compromises the health, naturalness and beauty of the river. Condominiums, hotels and office buildings generate more direct tax revenue than a park or a boat launch or a view. Such development could wall off views and restrict public access. Several office towers and condominium complexes have been constructed in recent years, but up to now, their locations have been in areas already heavily developed. That could change.

The ambitious 2012 Riverfront Plan and its follow-up will be a measure of the value placed on the river and the extent to which the modern river ethos prevails. While the plan was officially adopted by the city in 2012, formal adoption does not guarantee implementation. Further implementation requires the investment of dollars, effort and political capital for which there are always competing demands. Its continued support, or lack thereof, will be a test of the commitment of city officials and the public that elects them.

In addition to these intrinsic dangers, there are more mundane threats from the "new users" of the river—those who use it as an amenity. Although

Removing graffiti is a continuing problem for park staff. *Photo by author.*

Unfortunately, "leave no trace" is not yet part of everyone's mindset. *Photo by author.*

James River Advisory Council and other organizations sponsor riverside cleanups and regrettably find plenty to pick up. *Courtesy of rich young.*

the river ethos, valuing and promoting the health and beauty of the river, is broadly based, it is by no means universal, nor has it spread to all users. Philip Riggan makes this clear in his web post "Top 10 Ways Richmonders Are Ruining the James."[252] Some of society's untoward behavior carries over into the park and onto the riverbanks, including graffiti, litter, drunken conduct, vandalism and theft, failure to pick up after dogs.

So, the new river ethos is not without threats, both fundamental and mundane. Still, much has been accomplished, and prospects are bright. The new value set, along with the trajectory of change already accomplished, provides a promising future for the James River in Richmond. There is no doubt, however, that a range of issues, controversies and decisions lie ahead.

A HOPEFUL FUTURE

In many respects, the aspirations of early river visionaries have been met and even exceeded, so it might appear that the big job of transforming the James River in Richmond has been accomplished. Over the last seventy years, emphasis has moved from industrial to recreational use of the river. For the most part, sewage no longer enters the river untreated, and biologically, the river is healthier than it has been in many generations. Major threats to the river like the proposed Riverside Expressway on the south bank and hydroelectric power plants have been removed. The James River Park System has been established, has grown over the years and is now protected by a conservation easement, so there is public access to the river, something that was not available sixty years ago. All of the dams on the river through Richmond have been breached or had a fishway installed, so the possibility of anadromous fish returning to traditional spawning grounds has been achieved. There has been a steady growth in the boating and fishing on the river, and businesses that support such activities have grown. The floodwall has been constructed, protecting businesses that once were vulnerable to flooding. The Canal Walk and downtown riverfront improvements have added to river access, led the way in providing a showcase for history and established the possibility for significant commercial and tourism development. The Virginia Capital Trail, much of it along the river, has been completed, walking and mountain biking trails permeate the shoreline on both sides of the river and Potterfield Bridge was constructed on the remains of Brown's Island Dam. Cultural and entertainment events are held regularly on Brown's Island and other locations along the restored canal and river, and steps have been taken to protect important views along the river, although no guarantees are in place.

While local initiatives and local leaders deserve recognition and celebration for the significant accomplishments achieved, the river's transformation also has been a reflection of national and international trends, events, and movements. Without these, the transformation of the river could not have been achieved. The environmental movement was a fundamental force that enabled and inspired local efforts. During this pivotal period, Rachel Carson published *Silent Spring* (1962), the National Environmental Policy Act was signed into law (1970), the first Earth Day was held (1970), the Environmental Protection Agency was established (1970) and the Clean Water Act was passed (1972).

The increase in leisure time and resources is another force that facilitated the transformation of the river. With leisure time available, the river became a draw that it had not been before. The manufacturers of "recreational toys" aided the attractiveness of the river. At mid-twentieth century, paddlers had little choice other than a seventeen-foot aluminum canoe. In the 1970s, the choices began to multiply at a rapid rate, and now there are paddlecraft of virtually every size, description, purpose and price. The trails, too, are supported by specifically designed trail bikes, and hiking, running and biking clubs help bring users to the trails along the river. The tourist industry and the city's interest in promoting it have added an incentive for improving the new uses of the river.

Recent trends and the framework of new attitudes and beliefs about the river promise a bright future. Although city officials, riverside business owners and river advocates sometimes look enviously at riverfronts in other cities, Richmond has unique features and opportunities. San Antonio's River Walk, for example, had been held up as the best model when Canal Walk was accomplished, but it is far from a perfect fit. San Antonio's "river" is much like a long, continuous canal; Richmond has a canal broken by topography into pieces that are difficult to connect for recreation and tourism purposes. What Richmond does have, that San Antonio does not, is a large, varied, sometimes unruly river *and* a canal. Richmond's approach is different: ignore neither the canal(s) and its potential nor the river and its possibilities.

Following what seems to be an established trajectory of change, a sketch of the future of the James River in Richmond might include the following trends and occurrences.

CANALS will continue to be an important part of the riverfront and perhaps will even increase in significance. Some locks might be returned to working order, and Great Ship Lock could be made operable again. Perhaps this time the Norfolk Southern bascule bridge that blocks entry into Richmond Dock will also be repaired. Imagine water taxi service from the new turning basin to Rocketts. Maybe the Kanawha Canal westward from Tredegar to Maymont or Pump House—or even Bosher's Dam—might be made navigable and open for recreational use.

WHITEWATER will continue to be a unique draw for this urban river, both for paddlers on their own and those guided by professional outfitters. Even without stepping foot in a boat, visitors will be able to enjoy "just above the water" views. Perhaps Bridge Park and Missing Link will be completed and along with T-Pott, Mayo's Bridge and Belle Isle, make a triple downtown river loop that would be one of the most iconic walkways on the East Coast.

The JAMES RIVER PARK likely will continue to grow both physically and in the number of visits. Almost every island will be part of the park along with several new locations on shore. With multi-jurisdictional collaboration (with or without a new governance structure), the park could extend west in Henrico County past Bosher's Dam and east to Goode's Creek. The growth in visitor numbers will level off, as it inevitably must either as the result of policy change or natural trends.

The DOWNRIVER TIDAL AREA, trends and aspirations suggest, likely will see increased attention and public access. The Virginia Capital Trail passes through the area, and planning for more public access is well underway. Much of the area will be park and visitor amenities. Along with steps to protect views, one can imagine restaurants, visits by small cruise ships, water taxis, trails on both Chapel and Mayo's Islands and a facelift for Manchester Dock and Ancarrow's Landing. An increased number and variety of events, like Parade of Lights, dragon boat and crew races can be expected. One can even imagine strange events like remote-control boat races and over-water drone combat competitions.

The FLOODWALL is here to stay, but perhaps a new floodwall enhancement project will be initiated, this time with ample funding and new opportunities. Maybe even an aesthetic fix for the south-side riprap will be found.

The HEALTH OF THE RIVER and the habitat it supports will continue to improve, barring some unexpected pollution occurrence. The city will continue to push forward with pollution controls and continue to be among the nation's leaders in state-of-the-art wastewater treatment facilities. The ultimate fate of American shad on the river is uncertain, but there is still hope and since the Bosher's Dam fishway works well for other species, perhaps it will ultimately do so for American shad, the nation's "founding fish."

PANORAMIC VIEWS will be the beneficiary of both careful planning and advocacy pressure preventing them from becoming victim to the continuing growth of the city. Perhaps new public or semipublic views will be opened.

The SHOWCASE FOR HISTORY along the river is likely to expand. Many ghosts along the river are waiting to be brought to life. Preservation of historic sites and structures is more problematic, but perhaps the pressure to preserve and the appeal of adaptive reuse will provide adequate protection.

Lurking over these speculative projections of current trends are four fundamental questions:

- While city government has significantly increased its support for the river as a natural resource, tourist attraction and amenity, the river must compete with a great many other needs—schools, streets, a new coliseum and much more. Will other priorities push the river to the side and deny it the resources needed?

- As a regional, not just city attraction, should the city find a way to make the James River Park System a multi-jurisdictional entity, thereby drawing resources from a larger base?
- Notwithstanding the strategies to address the issue in the new park master plan, is the James River Park System on an internal collision course between its two fundamental goals: make the river accessible to the public and maintain an urban wilderness?
- Of the dozens of ideas and proposals in the 2012 Riverfront Plan, the recently adopted James River Park Master Plan, Venture Richmond's canal aspirations and other ideas, which should take priority? Which, if any, should be abandoned?

Recent trends and these few thoughts about the future imply a very busy river. Many river advocates value the little bit of wilderness that the river and its park-like surroundings provide and sometimes think that the river is best when it feels like it is one's own private playground. But to be preserved, that little bit of wilderness must be shared and involvement encouraged. Without sharing, which does take away some of the wilderness feel, it will no longer have the support and funding required to sustain it. It will be used for something else, and the wilderness will be gone. So, despite the irony, sharing is essential. That sharing requires some of the things river enthusiasts sometimes resist: planning, politics, management, rules, authority.

Many local individuals and organizations took the initiative, and sometimes fought political battles to bring about the achievements the transition from sewer to park required. City government, sometimes grudgingly and sometimes enthusiastically, has moved forward with its attention to the river. A variety of organizations continue the work of promoting the health and beauty of the river: the Falls of the James Scenic River Advisory Committee, the James River Association, Scenic Virginia, the James River Advisory Council, the James River Outdoor Coalition, rvaMORE, Richmond Sports Backers, Friends of James River Park, Enrichmond Foundation, Virginia Canals & Navigation Society, Capital Region Land Conservancy, Audubon Society, Sierra Club, Float Fishermen of Virginia, Coastal Canoeists and more. Critically important, city government now is supportive of the river in ways that it was not in decades past.

While the ultimate state of the river has not been reached, it seems likely that the framework for the foreseeable future of the James River in

Richmond has been set; hopefully, the new river ethos is robust enough to prevent any major deviations. Still, threats do exist. What is certain is that there will be work to do and issues to address; there will be conflict, but there also will be a sense of commitment, cooperation and passion among many who have come to consider the James River in Richmond their own.

NOTES

Chapter 2

1. A point made by Robert Steidel, formerly director of the Richmond Department of Public Utilities, is that sanitary engineering professionals of an earlier day did not simply consider the river as a vehicle for waste conveyance but rather as a sewage treatment facility. The river not only carried waste away and diluted it but through natural processes of decay also "treated" the sewage.
2. *Richmond Times-Dispatch*, January 9, 1949.
3. Quoted in Ann Woodlief, *In River Time: The Way of the James* (Chapel Hill, NC: Algonquin Books of Chapel Hill, 1985), 159, among other sources.
4. *An Environmental History: Stories of Stewardship in Virginia* (Virginia Department of Environmental Quality, April 2008), 6.
5. *Richmond Times-Dispatch*, August 22, 1949.
6. *Richmond News Leader*, September 9, 1952.
7. *Richmond News Leader*, August 13, 1952.
8. *Richmond News Leader*, September 10, 1952.
9. *Richmond News Leader*, October 8, 1963.
10. Margaret T. Peters, *Richmond Department of Public Utilities: 175 Years of Service* (Richmond, VA: City of Richmond Department of Public Utilities, 2008), 83.
11. Peters, *Richmond Department of Public Utilities*, 57, 66.
12. *Richmond Times-Dispatch*, August 31, 1971.

13. *Richmond Times-Dispatch*, December 4, 1977.

14. Newton H. Ancarrow, "Comments on the City of Richmond, Virginia's Environmental Assessment Statement and Federal Grant Application (C510-45201) for the Shockoe Combined Sewer Retention Basin," June 12, 1974, 14–15.

15. Peters, *Richmond Department of Public Utilities*, 84.

16. Scott Bass, "At First Flush," *Style Weekly*, January 13, 2010.

17. *Richmond News Leader*, December 6, 1983.

18. *Richmond Times-Dispatch*, June 2, 1985.

19. City of Richmond Department of Public Utilities, *Combined Sewer Overflow Study: Final Report*, October 1988.

20. Department of Public Utilities, *Overflow Study*, 1988, 120.

21. Ibid., 130.

22. The Falls of the James Scenic River Advisory Committee at times during its history was referred to as a board and at other times as a committee. For consistency, committee is used throughout.

23. Falls of the James Scenic River Advisory Committee, "Comments on Interim report VI—Environmental Information Document," 1989, 2.

24. *Richmond Time-Dispatch*, March 18 and 20, 1990.

25. *Richmond News Leader*, October 31, 1990.

26. *Richmond News Leader*, November 5, 1991.

27. *Richmond News Leader*, November 7, 1991; *Richmond Times-Dispatch*, November 8, 1991.

28. *Richmond Times-Dispatch*, December 10, 1991.

29. *Richmond Times-Dispatch*, January 7, 1992.

30. Peters, *Richmond Department of Public Utilities*, 118.

31. Michelle B. Kaszuba, "Combined Sewer Overflow" (master's thesis, Virginia Commonwealth University, 2002), 62.

32. Peters, *Richmond Department of Public Utilities*, 121.

33. Ibid., 120.

34. RVAH2O, www.rvah2o.org.

35. Reedy Creek Coalition, www.reedycreekcoalition.org.

36. *Richmond Times-Dispatch*, February 15, 1995.

37. *Richmond Times-Dispatch*, August 25, 2014.

38. *Richmond Times-Dispatch*, July 24, 2015.

Chapter 3

39. "Parkway" and "expressway" are used interchangeably here as they were during the period described.
40. J.J. Jewett to Ruffin Bailey with copies to John W. Pearsall and others, February 10, 1966.
41. John W. Pearsall to Richmond Metropolitan Authority, July 16, 1966.
42. John W. Pearsall III to Neighbor, October 27, 1966.
43. Jack Pearsall, personal communication, November 30, 2015.
44. G.C. Budd Corporation document November 29, 1966
45. Pearsall communication, November 30, 2015.
46. John W. Pearsall to Richmond Metropolitan Authority, December 14, 1966.
47. Statement by Charles A. Taylor, chairman, Richmond Metropolitan Authority, December 15, 1966, 1–2.
48. Taylor statement, December 15, 1966, 6.
49. *Richmond News Leader*, September 21, 1967.
50. *Richmond Times-Dispatch*, March 25, 1967.
51. "History of the Falls of the James Scenic River Advisory Board," unpublished document, January 30, 1995.
52. Louise Burke conversation, September 22, 1999.
53. Ibid.; "History of the Falls," 1995.
54. *Free-Lance Star* (Fredericksburg, VA), June 23, 1970.
55. Louise Burke to Bob Shaefer, July 14, 1970.
56. Burke conversation, September 22, 1999.
57. R.B. Young conversation, October 7,1999.
58. This and quotations that follow are from the transcript of the meeting of the Richmond Scenic James Council with Richmond City Council on November 17, 1970.
59. In attendance were representatives from the Richmond Metropolitan Authority: Charles Taylor, Jack Brent; from the city: Mayor Bliley, Conard Maddox, city attorney, James Carpenter, member of council, Alan Kiepper, city manager; and from the Richmond Scenic James Council: Allen Cripe, Richard Obenshain and R.B. Young.
60. Richard D. Obenshain to the Honorable Thomas J. Bliley Jr., December 3, 1970.
61. Jack Pearsall communication, November 30, 2015.

Chapter 4

62. *Richmond Times-Dispatch*, June 2, 1970.
63. *Richmond Times-Dispatch*, February 15, 1995.
64. *Richmond News Leader*, October 12, 1970. Bemiss also said the next step is to establish a Capital Region Park Authority with Richmond, Henrico, Chesterfield, a recommendation that has been made from time to time over the years but never implemented.
65. *Richmond Times-Dispatch*, February 8, 2015.
66. Quoted in *Richmond Times-Dispatch*, June 19, 1949.
67. *Richmond Times-Dispatch*, June 4, 1966.
68. *Richmond News Leader*, May 5, 1967; *Richmond Times-Dispatch*, July 4, 1968.
69. Stanley W. Abbott, "James River Park Development Narrative Report," 1968.
70. Stanley W. Abott to Jesse Reynolds, January 21, 1968.
71. Abbott, "Narrative Report," 1968, 2.
72. Abbott to Reynolds, January 21, 1968.
73. *Richmond News Leader*, February 20, 1973.
74. Joseph J. Shoman, "The James River Nature Park," September 1973, 10.
75. Ibid., 7.
76. *Richmond News Leader*, March 18, 1970.
77. *Richmond News Leader*, April 26, 1972.
78. City Staff, "Analysis of Proposed Development of 'Stern Property,'" May 11, 1972.
79. J. Robert Hicks, "Presentation to City Council," July 14, 1980.
80. Conversation with Ralph White, January 22, 2016.
81. R.B. Young to Mayor Thomas J. Bliley, March 30, 1972.
82. Carlton Abbot and Partners, "Proposed Alignment Studies for Riverside Trail Richmond, Virginia," June 5, 1992.
83. Presentation by R.B. Young before city council budget hearing, May 7, 1975.
84. *Richmond Times-Dispatch*, March 18, 1980.
85. Dan Caston, "River Access: Perspectives from a Paddler," *James River Reach*, Fall 2001.
86. *Richmond Times-Dispatch*, May 29, 2009.
87. *Richmond Times-Dispatch*, October 21, 2005.
88. *Richmond Times-Dispatch*, May 29, 2009.

Chapter 5

89. Steven M. Atran, Joseph G. Loesch, William H. Kriete Jr. and Ben Rizzo, *Feasibility Study of Fish Passage Facilities in the James River, Richmond, Virginia* (Virginia Commission of Game and Inland Fisheries, 1983), 19.

90. W.E. Trout III, James Moore III and George D. Rawls, *Falls of the James Atlas*, 2nd ed. (Virginia Canals and Navigations Society, 1995), 41.

91. Ibid., 31.

92. Atran et al., *Feasibility Study*, 4–5; Trout et al., *Falls of the James Atlas*, 12–14.

93. Atran et al., *Feasibility Study*, 3.

94. Ibid., 1983.

95. R.B. Young to Bud Bristow, May 31, 1991; Department of Game and Inland Fisheries, *Interim Report: Williams Island Dam Fish Passage and Safety Modifications*, February 1992.

96. Price Smith to Frank Harksen, November 6, 1991.

97. Charles T. Peters Jr. to David Whitehurst, February 19, 1993.

98. James River Association, *Tidings*, Winter 1992, 1.

99. *Richmond Times-Dispatch*, March 7, 1993.

100. John M. Mudre, John J. Ney and Richard J. Neves, *Analysis of Impediments to Spawning Migrations of Anadromous Fishes in Virginia Rivers* (Virginia Highway Research Council and Virginia Department of Highways and Transportation, 1985).

101. James River Association, *Annual Report: 1997–98*.

102. Alan Weaver, personal communication, August 28, 2019.

103. *Richmond Times-Dispatch*, April 21, 1999.

104. *Richmond Times-Dispatch*, May 25, 2009.

105. Department of Game & Inland Fisheries, "On the Road to Recovery: American Shad Restoration," www.dgif.virginia.gov/fishing/shad-restoration.

106. Quoted in *Richmond Times-Dispatch*, September 6, 2015.

107. VCU Rice Rivers Center, "Atlantic Sturgeon Restoration," https://ricerivers.vcu.edu/research/atlantic-sturgeon-restoration.

108. James River Association, *Tidings*, March 2014, 3.

109. NOAA, Chesapeake Bay Office, "Invasive Catfish," https://chesapeakebay.noaa.gov/fish-facts/invasive-catfish.

110. Chesapeake Bay Program, "Blue Catfish: *Ictalurus furcatus*," https://www.chesapeakebay.net/S=0/fieldguide/critter/blue_catfish.

111. "Invasive Catfish."

112. U.S. Fish and Wildlife Service, Draft Environmental Impact Statement: Resident Canada Goose Management, February/2002, I-2.

113. Ibid.
114. *Richmond Times-Dispatch*, March 25, 2002.
115. *Richmond Times-Dispatch*, January 16, 2012.
116. *Richmond Times-Dispatch*, April 29, 2004.
117. Ibid.

Chapter 6

118. *4ᵗʰ District Update*, Spring 2015.
119. W. Dudley Powers, May 25, 1920 quoted in Trout et al., *Falls of the James Atlas*, 32.
120. Account provided to the author by Greg Velzy.
121. *Richmond Times-Dispatch*, November 1, 1975; *Richmond News Leader*, November 3, 1975.
122. Stuart Bateman conversation, October 19, 1999.
123. Request for Proposals 8Z037, October 1997.
124. Request for Proposals, Z00043, August 23, 1999.
125. Addendum no. 1, attachment 2.
126. John Bryan, *The James River in Richmond: Your Guide to Enjoying America's Best Urban Waterway* (Richmond, VA: Charles Creek Publishing, 1997), 99.
127. James River Association website, www.thejamesriver.org.
128. Lisa Lambrecht, quoted in "10 Reasons We Bike James River Park Trail," *Richmond Times-Dispatch*, September 21, 2012, richmond.com.

Chapter 7

129. *Richmond Times-Dispatch*, April 18, 1993.
130. City of Richmond Planning Commission, *Richmond & the James: A Plan for Conservation, Recreation, Beautification*, 1967.
131. Unpublished paper, November 15, 1966.
132. James River and Kanawha Canal Parks Inc., unpublished paper, May 2, 1973.
133. Quotations from *Richmond Times-Dispatch*, October 19, 1969.
134. *Richmond Times-Dispatch*, January 10, 1971.
135. Document from James River and Kanawha Canal Parks, Inc. May 1, 1972.
136. Jack Pearsall conversation; Brenton S. Halsey, *Riverfront Renaissance* (Manakin-Sabot, VA: Dementi Milestone Publishing, 2016), 15–18.

137. *Richmond Times-Dispatch*, April 6, 1974.

138. Jack Pearsall conversation.

139. Paul A. Murphy, "The Kanawha Canal: Linking the Atlantic Ocean with the Ohio River," 1971, a pamphlet reprinted by the National Trust for Historic Preservation from *Historic Preservation*, 23. no. 3 (July–September 1971).

140. INTERPLAN/Danadjieva & Koenig Associates, *Vanishing & Returning Gardens of Richmond: A Development Program for the Riverfront*, prepared for the City of Richmond, September 1977.

141. Glave Newman Anderson & Associates, *Riverfront Flood Protection and Development Study*, April 1979.

142. Carlton Abbott and Partners, *The Richmond Canals Plan: A Conceptual Master Plan for the Restoration and Development of Richmond's Historic Canals* (Historic Richmond Foundation, 1988).

143. Halsey, *Riverfront Renaissance*, 31.

144. Young to Gottwald, May 9, 1989.

145. *Richmond News Leader*, April 4, 1989.

146. Deona Houff, "Troubled Waters: The Fight Over the Canal," *Style Weekly*, July 11, 1989.

147. *Richmond Times-Dispatch*, November 11, 1989.

Chapter 8

148. Virginius Dabney, *Richmond: The Story of a City* (Garden City, NY: Doubleday & Company, 1976), 22.

149. Earl Swift, *Journey on the James: Three Weeks Through the Heart of Virginia* (Charlottesville: University Press of Virginia, 2001), 107.

150. Ibid., 106.

151. Jay Harris, Dave Knapp and James Berry, *Hurricane Agnes… The Richmond Flood* (Lubbock, TX: C.F Boone, 1972), 7.

152. Ibid., 3.

153. *Richmond Times-Dispatch*, July 2, 1972.

154. U.S. Army Corps of Engineers, *James River Basin, Virginia Feasibility Report for Flood Control at Richmond*, October 1974.

155. *Richmond News Leader*, October 7, 1972.

156. *Richmond Times-Dispatch*, October 8, 1972.

157. U.S. Army Corps of Engineers, *James River Basin*, October 1974, Appendixes A-2.

158. Ibid., 28–29.

159. Ibid., 37.

160. Ibid., 33.

161. Ibid., 38.

162. Office of the Chief of Engineers, Department of the Army, *Final Environmental Impact Statement: Flood Protection Measures at Richmond, Virginia,* November 1975, 5.

163. *Revitalization News* 6, no. 7 (Fall 1989): 2.

164. Ibid., 2–3.

165. *Revitalization News* 2, no.1 (January 1984): 4.

166. *Richmond News Leader*, November 7, 1985.

167. *Richmond Times-Dispatch*, December 31, 1973.

168. Letter to Norfolk District, Corps of Engineers, March 25, 1974.

169. Conservation Council of Virginia Inc., "Position on Flood Protection for Richmond," September 22, 1973.

170. *Richmond News Leader*, August 13, 1980.

171. *Richmond News Leader*, May 28, 1986.

172. *Richmond Times-Dispatch*, September 28, 1980.

173. From printed program dedicating the Floodwall, October 21, 1994.

174. Carlton Abbott and Partners, *Richmond Floodwall Enhancement for the Proposed Richmond Floodwall Project*, April 1991.

175. *Richmond Times-Dispatch*, June 4, 1991.

176. Halsey, *Riverfront Renaissance*, 33.

177. Mary Jane Hogue, presentation at the dedication of restored canal, 1999.

178. Walter S. Griggs Jr., *Historic Disasters of Richmond* (Charleston, SC: The History Press, 2016), 22.

Chapter 9

179. James River Discovery Program Document, April 29, 1985.

180. Halsey, *Riverfront Renaissance*, 36.

181. *Richmond Times-Dispatch*, May 4, 1991.

182. Brenton Halsey, "Riverfront Effort Unites City's Past and Economic Future," *Richmond Times-Dispatch*, April 18, 1993.

183. Ibid., italics in original.

184. Suzan K. Cecil to Virginia Marine Resources Commission, November 10, 1995.

185. Richmond Riverfront Development Corporation, *Richmond Riverfront Master Development Plan*, July 1993.

186. Anne Jordan, "River of Dreams," *Governing*, August 1997, 28.

187. Halsey, *Riverfront Renaissance*, 43.

188. *Richmond Times-Dispatch*, February 27, 1999.

189. *Richmond Times-Dispatch*, April 21, 1999.

190. Halsey, *Riverfront Renaissance*, 80–83.

191. *Richmond Times-Dispatch*, September 6, 1996.

192. *Richmond Times-Dispatch*, August 24, 1993.

193. *Richmond Times-Dispatch*, January 25, 1997.

194. *Richmond Times-Dispatch*, June 1, 1999.

195. Ibid.

196. Ibid.

197. Halsey, *Riverfront Renaissance*, 133–46.

198. *Richmond Times-Dispatch*, May 6, 2008.

199. *Richmond Times-Dispatch*, July 28, 2009.

200. Mark Olinger presentation, February 18, 2014.

201. Attributed to Tyler Potterfield in Max Hepp-Buchanan, "The Rise of the Brown's Island Dam Walk," January 22, 2014, www.richmondoutside.com.

202. *Richmond Times-Dispatch*, August 25, 2017.

Chapter 10

203. T. Tyler Potterfield, *Nonesuch Place: A History of the Richmond Landscape* (Charleston, SC: The History Press, 2009), chapter 1.

204. Quoted by Potterfield, *Nonesuch Place*, 21.

205. Email from Gayla Mills, co-chair, Citizens Advocating Responsible Riverfront Development (CARRD) to email list, August 29, 2001.

206. Dominion Press Release, September 6, 2001.

207. CARRD information packet, September 18, 2001.

208. *Richmond Times-Dispatch*, September 13, 2001.

209. *Richmond Times-Dispatch*, October 2, 2001.

210. *Richmond Times-Dispatch*, November 6, 2001.

211. *Richmond Times-Dispatch*, November 26, 2001.

212. *Richmond Times-Dispatch*, November, 27, 2001.

213. *Style Weekly*, December 11, 2001.

214. *Richmond Times-Dispatch*, April 13, 2002.

215. Resolution No. 2002-R105, adopted April 22, 2002.

216. *Richmond Times-Dispatch*, June 18, 2002.

217. Ordinance Numbers: 2002-162, conditional zoning; 2002-163, RF-1; and 2002-164, RF-2.

218. Scenic Virginia press release, October 26, 2006.

219. Virginia Chapter of the American Society of Landscape Architects, *Richmond Riverfront Viewshed Study*, 2013–2014, 25.

220. *Richmond Times-Dispatch*, September 2, 2012.

221. Leighton Powell, personal communication.

Chapter 11

222. Trout et al., *Falls of the James Atlas*, 52.

223. Robert C. Layton, *Discovering Richmond Monuments: A History of River City Landmarks Beyond the Avenue* (Charleston, SC: The History Press, 2013), 99–100.

224. Trout et al., *Falls of the James Atlas*, 44.

225. Dabney, *Richmond*, 156; Layton, *Discovering Richmond Monuments*, 40–41.

226. Rebecca Bridges Watts, *Contemporary Southern Identity: Community Through Controversy* (Jackson: University Press of Mississippi, 2008), 73–86.

227. *Richmond Times-Dispatch*, November 7, 1999.

228. *Richmond Times-Dispatch*, July 14, 2019

229. Dabney, *Richmond*, 1.

230. Layton, *Discovering Richmond Monuments*, 19–20.

231. Ralph White, *An Interpretive Guide to Belle Isle* (Richmond, VA: James River Park System, 2003), 1.

232. "Belle Isle: Paradise and Pain," *Richmond Journal of History and Architecture* 1, no. 2 (Autumn 1994): 4.

233. Dabney, *Richmond*, 137.

234. Mary H. Mitchell, *Hollywood Cemetery: The History of a Southern Shrine* (Richmond: Virginia State Library, 1985), 7.

235. Trout et al., *Falls of the James Atlas*, 20.

236. Ibid., 16.

237. Harry M. Ward and Harold E. Greer Jr., *Richmond During the Revolution 1775–83* (Charlottesville: University Press of Virginia, 1977), 137.

238. Trout et al., *Falls of the James Atlas*, 14.

239. *Richmond Times-Dispatch*, June 30, 1991,

240. Harry Kollatz Jr., *True Richmond Stories: Historic Tales from Virginia's Capital* (Charleston, SC: The History Press, 2007), 148–49.

241. *Richmond Times-Dispatch*, December 4, 1977; June 30, 1991.

242. The Historical Marker Database.

243. Maureen Egan, *Insiders' Guide to Richmond, VA* (Guilford, CT: Globe Pequot Press, 2010), 172.

244. Benjamin Campbell, *Richmond's Unhealed History*. (Richmond, VA: Brandylane Publishers, 2012), 109.

245. Dabney, *Richmond*, 44–45, 279; Trout et al., *Falls of the James Atlas*, 51.

246. Trout et al., *Falls of the James Atlas*, 41.

Chapter 12

247. Ann M. Woodlief and Lynn D. Nelson, eds., *The James River as Common Wealth* (Notre Dame, IN: Foundations Press, 1984), 17–18.

248. Ibid., 19.

249. Ibid., 20.

250. Ibid., 20.

251. Young letter, July 18, 1974.

252. Posted March 30, 2012.

SELECTED BIBLIOGRAPHY

Bryan, John. *The James River in Richmond: Your Guide to Enjoying America's Best Urban Waterway*. Richmond, VA: Charles Creek Publishing, 1997.

Canal Committee of the Historic Richmond Foundation. *Richmond's Historic Waterfront: 1607–1865*. Richmond, VA: Historic Richmond Foundation, 1989.

Daniel, Will. *James River Reflections*. Atglen, PA: Schiffer Publishing, 2011.

Deans, Bob. *The River Where America Began: A Journey Along the James*. Lanham, MD: Rowman & Littlefield Publishers, 2007.

Halsey, Brenton S. *Riverfront Renaissance*. Manakin-Sabot, VA: Dementi Milestone Publishing, 2016.

Harris, Jay, Dave Knapp and James Berry. *The Richmond Flood*. Lubbock, TX: C.F. Boone Publisher, 1972.

Niles, Blair. *The James*. New York: Farrar & Rinehart, Inc., 1939. Expanded and published in 1945 as *The James: From Iron Gate to the Sea*.

Potterfield, T. Tyler. *Nonesuch Place: A History of the Richmond Landscape*. Charleston, SC: The History Press, 2009.

Ryan, David D. *The Falls of the James*. Richmond, VA: William Byrd Press, 1975.

Swift, Earl. *Journey on the James: Three Weeks through the Heart of Virginia*. Charlottesville: University Press of Virginia, 2001.

Totty, Dale. *Maritime Richmond*. Charleston, SC: Arcadia Publishing, 2004.

Trout, W.E., III, James Moore III and George D. Rawls. *Falls of the James Atlas*. 2nd ed. Lexington, VA: Virginia Canals and Navigations Society, 1995.

Woodlief, Ann. *In River Time: The Way of the James*. Chapel Hill, NC: Algonquin Books of Chapel Hill, 1985.

INDEX

Viewing rivers for over seventy years.

Visit us at
www.historypress.com

ABOUT THE AUTHOR

 alph Hambrick is a member and former chair of the Falls of the James Scenic River Advisory Committee and a member and former co-chair of the James River Advisory Council. He earned a bachelor of arts degree from Dartmouth College and a doctorate from Syracuse University and is professor emeritus in public policy and administration at Virginia Commonwealth University. He is a former whitewater canoe instructor, a raft guide and an all-around river enjoyer who does his writing from a home office overlooking the James River. Ralph and his wife, Linda, divide their time between Richmond, Virginia, and Hilltop Lakes, Texas.